PROBLE[M]
LEARNING
IN THE
EARTH AND SPACE
SCIENCE
CLASSROOM
K–12

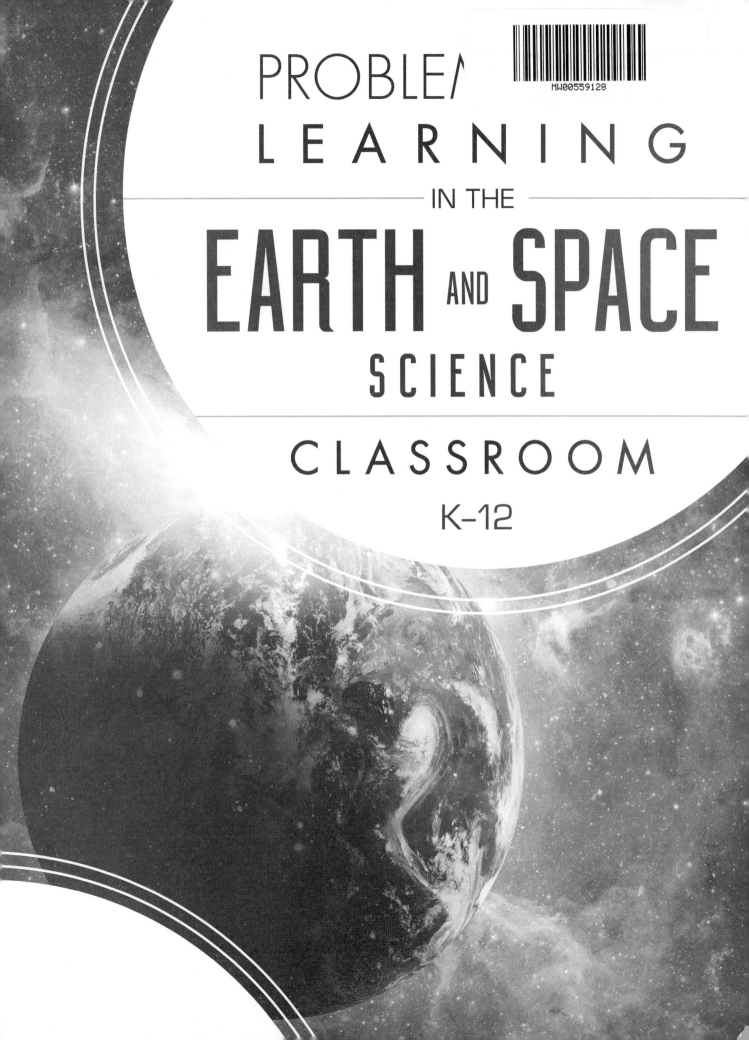

PROBLEM-BASED LEARNING

IN THE

EARTH AND SPACE

SCIENCE

CLASSROOM

K–12

TOM J. MCCONNELL · JOYCE PARKER · JANET EBERHARDT

National Science Teachers Association

Arlington, Virginia

Claire Reinburg, Director
Wendy Rubin, Managing Editor
Rachel Ledbetter, Associate Editor
Amanda Van Beuren, Associate Editor
Donna Yudkin, Book Acquisitions Coordinator

ART AND DESIGN
Will Thomas Jr., Director
Himabindu Bichali, Graphic Designer, cover and
 interior design

PRINTING AND PRODUCTION
Catherine Lorrain, Director

NATIONAL SCIENCE TEACHERS ASSOCIATION
David L. Evans, Executive Director
David Beacom, Publisher

1840 Wilson Blvd., Arlington, VA 22201
www.nsta.org/store
For customer service inquiries, please call 800-277-5300.

NSTA is committed to publishing material that promotes the best in inquiry-based science education. However, conditions of actual use may vary, and the safety procedures and practices described in this book are intended to serve only as a guide. Additional precautionary measures may be required. NSTA and the authors do not warrant or represent that the procedures and practices in this book meet any safety code or standard of federal, state, or local regulations. NSTA and the authors disclaim any liability for personal injury or damage to property arising out of or relating to the use of this book, including any of the recommendations, instructions, or materials contained therein.

Library of Congress Cataloging-in-Publication Data
Names: McConnell, Tom J., 1962- | Parker, Joyce, 1955- | Eberhardt, Janet,   1952-
Title: Problem-based learning in the earth and space science classroom, K-12 / Tom J. McConnell, Joyce Parker, Janet Eberhardt.
Description: Arlington, VA : National Science Teachers Association, [2017] | Includes index.
Identifiers: LCCN 2016043800 | ISBN 9781941316191 (print)
Subjects: LCSH: Earth sciences--Study and teaching (Elementary)--United States. | Earth sciences--Study and teaching (Secondary)--United States.
Classification: LCC QE47.A1 M33 2017 | DDC 550.71/273--dc23 LC record available at *https://lccn.loc.gov/2016043800*
eISBN: 978-1-941316-72-6

CONTENTS

PREFACE

In science education, there are numerous strategies designed to promote learners' ability to apply science understanding to authentic situations and build connections between concepts (Bybee, Powell, and Trowbridge 2008). Problem-based learning (PBL) (Delisle 1997; Gijbels et al. 2005; Torp and Sage 2002) is one of these strategies. PBL originated as a teaching model in medical schools (Barrows 1986; Schmidt 1983) and is relevant for a wide variety of subjects. Science education, in particular, lends itself to the PBL structure because of the many authentic problems that reflect concepts included in state science standards and the *Next Generation Science Standards* (*NGSS*; NGSS Lead States 2013).

The Problem-Based Learning Framework

PBL is a teaching strategy built on a constructivist epistemology (Savery and Duffy 1995) that presents learners with authentic and rich, but incompletely defined, scenarios. These "problems" represent science as it appears in the real world, giving learners a reason to collaborate with others to analyze the problem, ask questions, pose hypotheses, identify information needed to solve the problem, and find information through literature searches and scientific investigations. The analysis process leads the learners to co-construct a proposed solution (Torp and Sage 2002).

One of the strengths of the PBL framework is that learners are active drivers of the learning process and can develop a deeper understanding of the concepts related to the problem starting from many different levels of prior understanding. PBL is an effective strategy for both novices and advanced learners. PBL is also flexible enough to be useful in nearly any science context.

One of the challenges for teachers and educational planners, though, is that implementing PBL for the classroom requires advance planning. An effective problem should be authentic, and the challenges presented in the problems need to be both structured and ill-defined to allow genuine and productive exploration by students. Dan Meyer (2010) suggested that these problems help students learn to be "patient problem solvers." For most instructors, getting started with PBL in the science classroom is easiest with existing problems. However, there are very few tested PBL problems available in print or on the internet. Valuable resources exist that describe in general what PBL is, how to develop lessons, and how PBL can help students, but curriculum resources are much harder to find.

In this book, we present a discussion of the PBL structure and its application for the K–12 science classroom. We also share a collection of PBL problems developed as part of the

Problem-Based Learning Project for Teachers (PBL Project), a National Science Foundation–funded professional development program that used the PBL framework to help teachers develop a deeper understanding of science concepts in eight different content strands (McConnell et al. 2008; McConnell, Parker, and Eberhardt 2013). Each content strand had a group of participants and facilitators who focused on specific concepts within one of the science disciplines, such as genetics, weather, or force and motion. The problems presented in this book were developed by content experts who facilitated the workshops and revised the problems over the course of four iterations of the workshops. Through our work to test and revise the problems, we have developed a structure for the written problem that we feel will help educators implement the plans in classrooms.

Because the problems have been tested with teachers, we have published research describing the effectiveness of the problems in influencing teachers' science content knowledge (McConnell, Parker, and Eberhardt 2013). The research revealed that individuals with very little familiarity with science concepts can learn new ideas using the PBL structure and that the same problem can also help experienced science learners with a high degree of prior knowledge to refine their understanding and learn to better explain the mechanisms for scientific phenomena.

Alignment With the *Next Generation Science Standards*

To ensure that the problems presented here are useful to science teachers, we have included information aligning the objectives and learning outcomes for each problem with the *NGSS* (NGSS Lead States 2013). The *NGSS* present performance expectations for science education that describe three intertwined dimensions of science learning: science and engineering practices (SEPs), disciplinary core ideas (DCIs), and crosscutting concepts (CCs). The *NGSS* emphasize learning outcomes in which students integrate the SEPs, DCIs, and the CCs in a seamless way, resulting in flexible and widely applicable understanding.

The learning targets for the PBL problems included in this book were originally written with attention to the science concepts—what the *NGSS* call DCIs. The aim of the PBL Project was to enhance teachers' knowledge of these core ideas. But implicit in the design of the PBL process is the need for learners to use the practices of science and make connections between concepts that reflect the CCs listed in the standards. PBL problems align well with the *NGSS* because these real-world situations present problems in a similar framework: SEPs, DCIs, and CCs are natural parts of the problems. We describe the alignment of the PBL problems with the *NGSS* in more detail in Chapter 2. As states begin to adopt these standards or adapt them into state standards, Chapter 2 should help teachers and teacher educators fit the problems within their local curricula.

Intended Audiences and Organization of the Book

As mentioned earlier, the PBL problems in this book have been shown to be effective learning tools for learners with differing levels of prior knowledge. Some of the teachers who participated in the PBL Project used problems from the workshops in their K–12 classrooms, and facilitators with the project have also incorporated problems from this collection into university courses.

Chapter 2 discusses the alignment of the PBL problems and analytical framework with the *NGSS*. Chapter 3 describes strategies for facilitating the PBL lessons. In Chapter 4, we share tips for the classroom teacher on grouping students, managing information, and assessing student learning during the PBL process.

Chapters 5–8 present the problems we have designed and tested. Each chapter includes problems from one content strand (Earth's landforms and water, rock cycle and plate tectonics, weather, or astronomy), alignment with the *NGSS*, the assessment questions we used to evaluate learning, model responses to the assessments, and resources for the teacher and students that help provide relevant information about the science concept and problem. To help you locate the problems that are most appropriate for your classroom, we have included a catalog of problems (see p. xi); the catalog is in tabular format and will let you scan the list of problems by content topic, keywords and concepts, and grade bands for which the problems were written.

We hope that this collection of problems will serve as a model for educators who want to design and develop problems of their own. For instance, there are problems included in this book that relate to the local landforms and examples that reflect contexts relevant to Michigan, where the PBL Project was located. A teacher in a place that does not share similar conditions may find that his or her students cannot relate to the scenario described in the problem. In these cases, we encourage teachers to modify and adapt problems to fit contexts familiar to their own students. Chapter 9 discusses features of an effective problem that can help guide the efforts of teachers wishing to create their own PBL lessons. As you modify and implement lessons from these books, you can begin to develop your own problems that meet the needs of your students.

This book is the second volume in a series; the first volume presented life science problems. We present Earth and space science problems in this volume, and we will offer problems specifically written for teaching physics in the next volume to be published. The fourth volume will contain tips and examples for planners of teacher professional development programs.

Safe and Ethical Practices in the Science Classroom

With hands-on, process- and inquiry-based laboratory or field activities, the teaching and learning of science today can be both effective and exciting. Successful science teaching needs to address potential safety issues. Throughout this book, safety precautions are described for investigations and need to be adopted and enforced in efforts to provide for a safer learning and teaching experience.

Additional applicable standard operating procedures can be found in the National Science Teachers Association's Safety in the Science Classroom, Laboratory, or Field Sites document (*www.nsta.org/docs/SafetyInTheScienceClassroomLabAndField.pdf*).

Disclaimer: The safety precautions of each activity are based in part on use of the recommended materials and instructions, legal safety standards, and better professional practices. Selection of alternative materials or procedures for these activities may jeopardize the level of safety and therefore is at the user's own risk.

References

Barrows, H. S. 1986. A taxonomy of problem-based learning methods. *Medical Education* 20 (6): 481–486.

Bybee, R. W., J. C. Powell, and L. W. Trowbridge. 2008. *Teaching secondary school science: Strategies for developing scientific literacy.* Upper Saddle River, NJ: Prentice Hall.

Delisle, R. 1997. *How to use problem-based learning in the classroom.* Alexandria, VA: Association for Supervision and Curriculum Development.

Gijbels, D., F. Dochy, P. Van den Bossche, and M. Segers. 2005. Effects of problem-based learning: A meta-analysis from the angle of assessment. *Review of Educational Research* 75 (1): 27–61.

McConnell, T. J., J. Eberhardt, J. M. Parker, J. C. Stanaway, M. A. Lundeberg, M. J. Koehler, M. Urban-Lurain, and PBL Project staff. 2008. The PBL Project for Teachers: Using problem-based learning to guide K–12 science teachers' professional learning. *MSTA Journal* 53 (1): 16–21.

McConnell, T. J., J. M. Parker, and J. Eberhardt. 2013. Problem-based learning as an effective strategy for science teacher professional development. *The Clearing House* 86 (6): 216–223.

Meyer, D. 2010. TED: Math class needs a makeover [video]. *www.ted.com/talks/dan_meyer_math_curriculum_makeover?language=en.*

NGSS Lead States. 2013. *Next Generation Science Standards: For states, by states.* Washington, DC: National Academies Press. *www.nextgenscience.org/next-generation-science-standards.*

Savery, J. R., and T. M. Duffy. 1995. Problem based learning: An instructional model and its constructivist framework. *Educational Technology* 35 (5): 31–38.

Schmidt, H. G. 1983. Problem-based learning: Rationale and description. *Medical Education* 17 (1): 11–16.

Torp, L., and S. Sage. 2002. *Problems as possibilities: Problem-based learning for K–16 education.* 2nd ed. Alexandria, VA: Association for Supervision and Curriculum Development.

CATALOG OF PROBLEMS

ACKNOWLEDGMENTS

We wish to thank the following individuals who helped design, revise, and facilitate the problem-based learning (PBL) lessons presented in this book. Their expertise and insight were instrumental in the development of the problems and tips on facilitating PBL learning.

Michigan State University

Diane Baclawski

Kazuya Fujita

Merle Heidemann

Susan Jackson

Edwin Loh

Christopher Reznich

Mary Jane Rice

Duncan Sibley

Lansing Community College

Alex Azima

K. Rodney Cranson

Christel Marschall

Dennis McGroarty

Teresa Schultz

Jeannine Stanaway

Mott Community College

Judith Ruddock

DeWitt High School

Mark Servis

Retired Faculty

Roberta Jacobowitz (Otto Middle School)

Barbara Neureither (Holt High School)

Zandy Zweering (Williamston Middle School)

Ingham Intermediate School District

Theron Blakeslee

Martha Couretas

ABOUT THE AUTHORS

Tom J. McConnell is an associate professor of science education in the Department of Biology at Ball State University, Muncie, Indiana. He teaches science teaching methods courses for elementary and secondary education majors and graduate students and a biology content course for elementary teachers. His research focuses on the impact of professional development on teacher learning and student achievement and on curriculum development for teacher education programs. Dr. McConnell is an active member of the Hoosier Association of Science Teachers and the National Association for Research in Science Teaching.

Joyce Parker is an assistant professor in the Department of Earth and Environmental Sciences at Michigan State University in East Lansing. She teaches a capstone course for prospective secondary science teachers. Her research focuses on student understanding of environmental issues. She is an active member of the National Association for Research in Science Teaching and the Michigan Science Teachers Association.

Janet Eberhardt is a retired teacher educator and assistant director emerita of the Division of Science and Mathematics Education at Michigan State University in East Lansing. She has served as a consultant with the Great Lakes Stewardship Initiative and Michigan Virtual University. Her work has focused on designing effective and meaningful teacher professional development in the areas of science and mathematics.

1

DESCRIBING THE PROBLEM-BASED LEARNING PROCESS

A s a science teacher, you probably use a variety of approaches and strategies in the classroom. On any given day, you may lecture, lead group discussions, teach an inquiry-based lab, assign projects, ask students to complete individual reading and writing assignments, and perform many other types of tasks. All of these strategies have a legitimate purpose, and we encourage teaching that employs a diverse range of activities.

Why Problem-Based Learning?

In this chapter, we will discuss why problem-based learning (PBL) is one of the many tools you should keep in your teaching toolbox, ready to be used at appropriate times during your teaching. We will also give you some background information about how PBL was developed, how it works in a range of disciplines, and a basic framework for a PBL lesson. In later chapters, we will provide further detail on the "nuts and bolts" of teaching a PBL lesson and how to develop and facilitate learning activities using this strategy. The advice we will offer and the science problems we will share in later chapters come from our own experiences in using PBL to teach concepts to teachers. Many of the lessons have also been used with students across a wide range of age groups.

The reason we have used these lessons is because of a driving philosophy that it is imperative to help students develop the ability to inquire, solve problems, and think critically and independently (Barell 2010). Many of the thinking skills directly taught in the PBL process are included in the goals of 21st-century skills (Barell 2010; Ravitz et al. 2012). PBL is well suited to achieving the goal of developing thinking skills because it presents learners with authentic stories that require application of scientific concepts to construct and evaluate possible actions. In the process of solving problems, students plan, gather, and synthesize information from multiple sources or findings from investigations, evaluate the credibility of their sources, and communicate their ideas as they justify their claims. Students are guided by a set of simple prompts that help them organize information and generate questions and hypotheses.

In our experience, learners quickly adopt this framework as a habit of mind, and they begin to apply this critical-thinking strategy to other problems and real-world situations. The framework becomes a habit because the process is easily internalized and uses simple language. Asking the question "What do we know?" is easy for most students to remember and use, and the rest of the framework is just as direct and intuitive. This process also resembles KWL (McAllister 1994), a formative assessment strategy used widely in elementary classrooms. In KWL, students are asked to verbalize and record a list of what they "Know," what they "Want" to know, and what they have "Learned." The feature added by the PBL framework that makes it so "scientific" is the inclusion of hypotheses, leading students to make predictions and justify them.

Teachers in the professional development program for which these problems were developed very quickly adopted the language and turned "PBL" into a verb. When they encountered new problems, they initiated the process with phrases like "Let's PBL this." K–12 students are just as quick to adopt the cognitive framework. This is one of the benefits of using PBL in your teaching.

Historical Background of PBL as a Process

PBL's origins are rooted in this same desire to help learners solve real-world problems. PBL was originally developed as a strategy for developing content knowledge in the context of assessing and diagnosing patients (Barrows 1980). Medical students had been very successful in memorizing information, but when asked to use the information to diagnose a patient, they were unable to apply their knowledge. What was lacking in their understanding was how the ideas they had memorized were useful in diagnosing and treating patients in an authentic "problem" they would encounter as doctors. The challenge to medical school faculty was finding a way to teach students to think like doctors, not like students preparing for a test. PBL presents opportunities in just such a contextualized manner, so medical schools began using this strategy. PBL was shown to be effective in both helping medical students learn anatomy, pathology, and medical procedures and applying this knowledge to medical cases. Thus, PBL became widespread in medical schools.

The same issues seen in the field of medical education are important concerns for science students, too. Just as second-year medical students struggle to transfer what they learn into practice, science students struggle to understand how discovering the pattern of changes in phases of the Moon is helpful in explaining real-world issues or how plate tectonics shapes the world in which they live. Bransford and Schwartz (1999) suggested that transfer of knowledge is enhanced if the concepts are shown in a variety of contexts, rather than always presenting them in the same or very similar contexts. They also recommended using metacognition to support the transfer of knowledge across contexts. One of the strengths of PBL is that the framework we will present is a metacognitive structure—students are expected to be aware of what they know and what they need to know to solve the problem.

Bringing PBL to Other Disciplines

Since its beginnings in medical education, PBL has been adapted to business, law, law enforcement, and other subjects (Hung, Jonassen, and Liu 2008) and has been modified for science teaching (Allen et al. 2003; Gordon et al. 2001). Research by Hmelo-Silver (2004) suggested that PBL leads to increased intrinsic motivation of learners to become more self-directed. Another study reported that teachers who use PBL in their classrooms teach more 21st-century skills (Ravitz et al. 2012).

In this book, the model presented for using PBL to teach science content has similar features to the PBL activities from other subjects, but it has been refined through research-based evaluation of the process when used for teaching science content in the PBL Project for Teachers (McConnell et al. 2008), as described in the next section.

The PBL Project for Teachers

The context in which the materials presented in this book were created was the PBL Project for Teachers, a National Science Foundation–funded teacher professional development program (McConnell et al. 2008).[1] The PBL Project for Teachers was designed to accomplish several goals, including deepening K–12 teachers' scientific understanding, developing inquiry-based science lesson plans, and facilitating a form of reflective practice that applied the same PBL principles to the study of teaching.

In this program, K–12 teachers spent three days of a two-week institute learning science content surrounding standards they had identified as areas of need in their curricula. Facilitators for each of eight content strands planned PBL lessons to address those specific standards. These facilitators were experts in their respective science content areas who worked in teams of at least three. The teams wrote PBL problems that addressed the science standards teachers identified. Then, the teams shared these problems with peers for review. The problems were then tested and revised in an iterative fashion over four cohorts of teachers. The final versions were the basis for the problems found in this book and in future volumes in the PBL series.

The activities were modified for use in the K–12 classroom, with a focus on problems for life science, Earth and space science, and physics. These modifications included changing the context of the story to relate more to students in specific grade bands and changing the reading level to match the target audience. The concepts addressed in the problems remained consistent, in part because pre-assessments with teachers revealed very similar prior understandings as K–12 students, especially for teachers who were not science majors. Research to assess content learning showed that most of the teachers gained a deeper understanding of their chosen content as a result of the PBL lessons (McConnell, Parker, and Eberhardt 2013). Participants then used the content knowledge they gained to develop inquiry-based lesson plans and used PBL to analyze problems in teaching practice. During this stage, many of the problems were tested in K–12 and college courses, with revisions to address any difficulties encountered.

---

[1] National Science Foundation special project number ESI-03533406, as part of the Teacher Professional Continuum program.

The PBL Framework

In this book we use the same framework for designing and facilitating the PBL lessons that we used throughout the PBL Project for Teachers. This framework draws from the guidelines described by Torp and Sage (2002) and the model used by the Michigan State University College of Human Medicine (Christopher Reznich, personal communication, October 11, 2004). In this model, students are presented with a problem, usually in the form of a story divided into two parts (Christopher Reznich, personal communication, October 11, 2004). There can be more than two parts to the story, but the key feature is that information is presented to students in stages. Figure 1.1 shows a representation of the PBL process.

Figure 1.1. The PBL Process

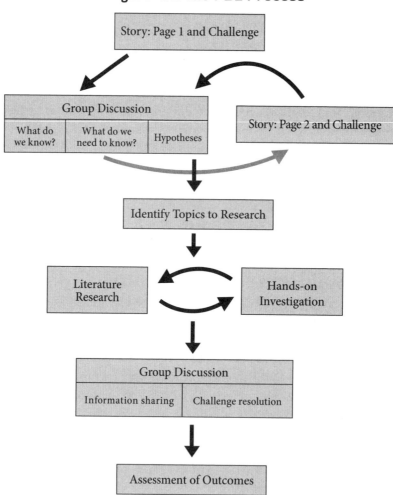

One of the keys to PBL is to develop a problem that is open-ended and ill-structured (Gallagher 1997; Torp and Sage 2002). By ill-structured, we mean a problem in which not all the information needed to solve the problem is presented to the learner, and some of the information presented may not be needed. In the real world, problems do not present themselves with a set number of variables or with a value for every variable provided but one. So the PBL lesson starts by presenting one of these scenarios and expecting students to unpack the problem and construct a path to solve the problem, much as Dan Meyer (2010) describes for teaching math when he cites Albert Einstein: "The formulation of a problem is often more essential than its solution, which may be merely a matter of mathematical or experimental skill." PBL offers a way to engage students in thinking about the problem, developing a strategy to solve the problem, and exploring the content knowledge needed to achieve a solution.

Torp and Sage (2002) also use language to describe the scenario presented by the teacher that is common to most literature on PBL. They refer to the story as the dilemma. We used the same language in the PBL Project for Teachers, but for this book series we have changed the terms. The origins of the word dilemma suggest that there are "two answers" to the scenario, but in reality a good *dilemma* is likely to have (at least at first glance) far more possible answers. So we have elected to call our scenarios *problems*. We will present the problems in the form of a story that ends in a challenge, which helps define the task students will take on and launches the research and analysis that follow. The challenge then serves as a focal point that defines what successful completion will require.

Structure of the Problem

In Chapters 5–8, we present PBL problems, which are the lessons that we have developed to share with teachers. Each chapter focuses on a content strand that fits within the Earth and space science discipline. These chapters begin with a description of the learning goals (Big Ideas) and the conceptual barriers students often face as they learn the concepts the problems address.

We have developed a consistent structure for each problem to help you find and use the resources you need. The main components of each problem are listed below and briefly described in the following subsections; see Chapter 3 for more details on facilitating each component:

- Overview and Alignment With the *NGSS*
- Page 1: The Story
- Page 2: More Information
- Page 3: Resources, Investigations, or Both (not included in every problem)

- Teacher Guide (with model response and activity guide)

- Assessment

Overview and Alignment With the *NGSS*

The first page of each problem provides information about the learning goals for the lesson in alignment with the *Next Generation Science Standards* (*NGSS*; NGSS Lead States 2013). The problems presented in this book were developed prior to the publication of the *NGSS*, but the nature of the PBL process lends itself to the type of three-dimensional learning described in the standards. Chapter 2 describes in more detail the alignment of the problems with the *NGSS* and the way we present the alignment in the tables that accompany each problem. The alignment table presented on the first page of each problem lists the performance expectations, science and engineering practices, disciplinary core ideas, and crosscutting concepts that are addressed in the lesson. Below the table is a list of keywords and concepts and a short description (under the heading "Problem Overview") of the context of the problem, which are intended to help teachers select problems that best suit their curricular needs.

We have also suggested some interdisciplinary connections for each chapter. For elementary and middle school teachers, it may be helpful to demonstrate to administrators that students are developing more than just science content during the lesson. Our lists are by no means exhaustive, so feel free to make other interdisciplinary connections as well.

Page 1: The Story

In this component, a very brief story presents an authentic scene that sets the context for the problem, followed by a "challenge" statement making it clear what the learners are expected to accomplish. In our model, Page 1 is the "engagement" activity to draw students into the learning situation and show them a real-world reason to know some science concepts. The Page 1 story ends with "Your Challenge," a question or series of questions that presents the students with a problem they need to resolve.

This page is the starting point to a whole-class discussion of the problem using a highly structured analytical framework. This framework prompts students to generate ideas and questions within three categories: "What do we know?" "What do we need to know?" and "Hypotheses." Table 1.1 gives an example of the framework as it might appear when used in the Leave It to the Masses problem in Chapter 7, "Weather." In Chapter 3, we present in more detail how the teacher facilitates this discussion, but one product of the initial discussion will be an analysis of the Page 1 story.

Page 2: More Information

During the discussion of the Page 1 story, students will eventually exhaust their ideas and questions using the analytical framework described in the preceding subsection. When the generation of ideas slows down, the facilitator presents Page 2. This part of the story

**Table 1.1. Sample of PBL Analytical Framework:
Leave It to the Masses Problem**

WHAT DO WE KNOW?	WHAT DO WE NEED TO KNOW?	HYPOTHESES?
• We will show four weather maps with temperature, wind, Doppler radar, and highs and lows. • Air mass is a body of air with similar temperature and moisture. • Warm air is less dense than cooler air. • Warm air has lower pressure than cooler air. • Fronts are boundaries between masses where weather is changing.	• What is Doppler radar? What can it detect? • What do air masses have to do with the map of winds? • How do you predict where air masses will be tomorrow? • What makes molecules in the air move? • How can drier air be denser than moist air? (Seems wet should be dense!)	• Cold, dry air will have high winds because it has more pressure and is denser. • Air masses will move from hot to cold because heat is transferred to cold things. • Air masses move west to east because that's the direction the wind usually blows.

is sometimes longer than Page 1, and it gives students more details about the story that are needed to solve the problem. There may also be new unnecessary information, but students will again decide what is important during a second round of the analytical discussion. During this discussion, students' ideas are added to the same three columns generated from discussion of Page 1. The challenge is repeated at the end of Page 2 to help the students keep that focus.

Page 3: Resources, Investigations, or Both

During the second round of the analytical discussion, groups also need to decide what their priorities will be for doing further research. The class needs to decide which items in the "What do we need to know?" column to explore further, then create a plan for finding information to address those items. The teacher can ask students to search for information individually or in groups, set time limits, plan for access to a computer lab or the technology in the classroom, and define a plan for sharing information when the searches are completed. In some cases, this research phase may also include some hands-on investigations or field observations.

As groups move on to some small-group research, they may need help finding relevant information. The Resources page includes references and links that may streamline the search process so that groups can make the best use of class time. The purpose of this list is to help the teacher steer students toward these resources during the research phase of the lesson. In some problems, we list resources on Page 2 (More Information) as well as on a separate Resources page. When we do, this is intended to push students to use those resources as they analyze the problem and scaffold their analysis process. To find the resources easily, you can view the Resources archive on the book's Extras page at *www. nsta.org/pbl-earth-space*.

Some problems include suggested investigations instead of or in addition to resources on Page 3. Each investigation provides a list of materials and a step-by-step procedure. Some investigations will have a "Safety Precautions" section to address safety standards and practices.

Once students have explored the resources or completed an investigation, the class convenes as a large group. In this final group discussion, students review the three columns in the analytical framework and present and discuss proposed solutions to the problem. There may be more than one appropriate solution to the problem, so students can justify and defend their ideas, using the evidence and resources they have collected to support their ideas.

Teacher Guide

The Teacher Guide includes the problem context, a model response, and, for some problems, an activity guide. This component of the PBL lesson provides resources for you, the teacher, to use. The "Problem Context" section gives you some background information about the scenario and science concepts associated with the problem, in case you are not familiar with the scenario. Each of the problems is aligned with standards, but this context goes a bit deeper and can help you explain the problem to students if needed. In some cases, this section also offers alternative contexts that may help you tailor the problem to your students.

The model response helps you assess the quality of responses your students generate. Just as you write down your ideal answer on an essay question before you grade it, we felt it would help to see what our content experts expect to see as a complete and accurate response. But keep in mind there may be more than one solution to the problem! The concepts included in our model responses could sometimes be applied in different and creative ways. Your students may not be able to provide explanations at this level, but knowing the explanation will help you assess how deeply your students understand the problem and its solution. Because of the wide variation in students' writing abilities and styles, we did not try to use student language in the model response. Instead, it is intended to give teachers an example of the concepts that are appropriate for an answer.

Some of the problems presented in Chapters 5–8 include hands-on activities that support students' learning. Some of these are inquiry activities, while others are models or examples to help illustrate concepts. For these lessons, we have included an "Activity

Guide" section with instructions on implementing the activity. These activity guides will resemble lesson plans you might find in other resources, with a list of materials, instructions for finding or constructing materials, and a description of the activity and the concepts it addresses. For example, in the Lassen's Lessons problem in Chapter 6, "Rock Cycle and Plate Tectonics," the "Activity Guide" section suggests having students make models of tectonic plate movement using graham crackers and frosting. The "Activity Guide" section may also include links to other resources or diagrams you may need to copy as a template for the materials you will make for the classroom. In some problems, the "Activity Guide" section will describe ways to incorporate the investigations listed on Page 3 of the problem into your lesson plans. A safety precautions note will be included if appropriate.

Assessment

We have provided assessment items in a variety of formats to help you gauge student learning. The assessment items include transfer tasks, solution summaries, two types of open-ended questions for pre/post assessment (general questions and application questions), and common beliefs inventories. For each problem in Chapters 5–8, model responses are provided for each type of assessment.

One of our goals is to help students apply concepts to new contexts. Transfer tasks can be used to assess that. Each transfer task is an open-response item in which the same concept addressed in the problem can be applied. For example, in the Keweenaw Rocks problem in Chapter 6, students learn about the geology of a region that was once a divergent plate boundary. The transfer task asks them to predict what types of rocks and formations they might expect to see in the East African Rift valley of Africa, another divergent plate boundary in a very different part of the world.

With each content strand, we have also included two types of open-ended assessment questions: general questions, which assess the breadth of knowledge about concepts, and application questions, which assess a student's ability to explain how the concepts apply to specific phenomena. In addition, there is a "Common Beliefs" assessment section with statements that reflect both accurate and inaccurate understandings; students are asked to answer *true* or *false* and to explain their reasoning with their answers.

Assessment is discussed in more detail in Chapter 4 ("Using Problems in K–12 Classrooms"), including the role of each type of assessment, the design of the assessments, and how they support teaching and learning in the PBL framework.

References

Allen, D. E., B. Duch, S. Groh, G. B. Watson, and H. B. White. 2003. Professional development of university professors: Case study from the University of Delaware. Paper presented at the international conference Docencia Universitaria en Tiempos de Cambio [University Teaching in Times of Change] at Pontificia Universidad Católica del Perú, Lima.

Barell, J. 2010. Problem-based learning: The foundation for 21st century skills. In *21st century skills: Rethinking how students learn*, eds. J. Bellanca and R. S. Brandt III, 175–199. Bloomington, IN: Solution Tree Press.

Barrows, H. S. 1980. *Problem-based learning: An approach to medical education.* New York: Springer.

Bransford, J. D., and D. L. Schwartz. 1999. Rethinking transfer: A simple proposal with multiple implications. *Review of Research in Education* 24 (1): 61–100.

Gallagher, S. A. 1997. Problem-based learning. *Journal for the Education of the Gifted* 20 (4): 332–362.

Gordon, P. R., A. M. Rogers, M. Comfort, N. Gavula, and B. P. McGee. 2001. A taste of problem-based learning increases achievement of urban minority middle-school students. *Educational Horizons* 79 (4): 171–175.

Hmelo-Silver, C. E. 2004. Problem-based learning: What and how do students learn? *Educational Psychology Review* 16 (3): 235–266.

Hung, W., D. H. Jonassen, and R. Liu. 2008. Problem-based learning. In *Handbook of research on educational communications and technology*, 3rd ed., eds. J. M. Spector, M. D. Merrill, J. van Merriënboer, and M. P. Driscoll, 485–506. New York: Routledge.

McAllister, P. J. 1994. Using KWL for informal assessment. *Reading Teacher* 47 (6): 510–511.

McConnell, T. J., J. Eberhardt, J. M. Parker, J. C. Stanaway, M. A. Lundeberg, M. J. Koehler, M. Urban-Lurain, and PBL Project staff. 2008. The PBL Project for Teachers: Using problem-based learning to guide K–12 science teachers' professional development. *MSTA Journal* 53 (1): 16–21.

McConnell, T. J., J. M. Parker, and J. Eberhardt. 2013. Problem-based learning as an effective strategy for science teacher professional development. *The Clearing House* 86 (6): 216–223.

Meyer, D. 2010. TED: Math class needs a makeover [video]. *www.ted.com/talks/dan_meyer_math_curriculum_makeover?language=en*.

NGSS Lead States. 2013. *Next Generation Science Standards: For states, by states.* Washington, DC: National Academies Press. *www.nextgenscience.org/next-generation-science-standards*.

Ravitz, J., N. Hixson, M. English, and J. Mergendoller. 2012. Using project based learning to teach 21st century skills: Findings from a statewide initiative. Paper presented at Annual Meeting of the American Educational Research Association, Vancouver, BC, Canada.

Torp, L., and S. Sage. 2002. *Problems as possibilities: Problem-based learning for K–16 education.* 2nd ed. Alexandria, VA: Association for Supervision and Curriculum Development.

2

ALIGNMENT WITH STANDARDS

One of the issues all teachers need to consider when designing curriculum and planning lessons is the standards for their content area. The content experts who wrote the problem-based learning (PBL) problems presented in this book were very cognizant of the importance of standards and used standards to guide the selection and creation of the topics, the tasks included in the lessons, and the assessments included with each problem.

In fact, the list of problems developed for each content strand was created to address concepts aligned with state and national standards (NRC 1996). Each participant in the four cohorts of teachers in the PBL Project for Teachers was asked to list the top three choices of standards they felt needed the most development in his or her own curriculum. From these lists, the planners developed content strands to address concepts that would meet each teacher's needs.

The standards used in this process included the Michigan Grade Level Content Expectations (MDOE 2007) and the *National Science Education Standards* (NRC 1996), both of which were the relevant standards at the time of the PBL Project. Since then, a new set of standards has been published. The standards alignment presented in this chapter addresses the *Next Generation Science Standards* (*NGSS*; NGSS Lead States 2013). As you use the problems in this book, you should also consider your state standards and local curriculum maps to guide your choices.

A Framework for K–12 Science Education and the NGSS

In 2012, the National Research Council released a document describing the structure of the new standards for science education, *A Framework for K–12 Science Education* (NRC 2012). The *Framework* laid out this structure as the foundation for the new standards that guide teachers to address more than just content. The PBL problems in Chapters 5–8 align well with the *Framework* because learners apply a variety of process skills and practices as they engage with specific science concepts, all of which fall within overarching themes that tie all the sciences together.

The *Framework* labels these different types of skills and concepts as three dimensions of science learning. Each of these dimensions is directly connected to helping students relate to and understand any scientific phenomenon. The term *dimensions* is meant to connote that the concepts and practices should be learned and used simultaneously rather than consecutively in the pursuit of understanding. Teachers should use phenomena (rather than generalities) as the context in which students practice and develop the three dimensions of scientific literacy. As shown in Box 2.1, the three dimensions of the *Framework* are science and engineering practices (SEPs), crosscutting concepts (CCs), and disciplinary core ideas (DCIs). In the next three subsections, we discuss the ways in which these dimensions of the *Framework* align with and are expressed in the PBL problems included in this book.

**Box 2.1. The Three Dimensions
of *A Framework for K–12
Science Education***

- Science and engineering practices

- Crosscutting concepts

- Disciplinary core ideas

Each of the problems in Chapters 5–8 begins with an overview section that includes a table outlining the problem's alignment with the *NGSS*. Each table includes the performance expectations, science and engineering practices (this is the term used in Appendix F of the *NGSS*), DCIs, and CCs associated with the problem. To help you find problems that fit your needs, we have included a catalog of problems (p. xi) that lets you see at a glance which problems will help you teach specific concepts within their specific content strands; the catalog includes keywords and concepts and the grade bands that include related standards.

Science and Engineering Practices

One of the dimensions in the *NGSS* is science and engineering practices (see Box 2.2). These practices describe skills and processes that scientists use in *doing* science, but they are more than just skills, so the authors of the *Framework* (NRC 2012) used the term *practices*. Most teachers will recognize these practices because the language incorporates the process skills and elements of the nature of science (Lederman 1999) expressed in the standards published by state departments of education. The SEPs included in Appendix F of the *NGSS* are shown in Box 2.2.

One of the benefits of PBL is that students are developing many of these practices as they progress through the analytical framework we introduced in Chapter 1 and explain in more detail in Chapter 3, "Facilitating Problem-Based Learning." If you present each problem in the format we have used, students take part in practices 1 and 4 during the discussion of the story; practices 3, 4, and 8 as they complete research on the problem; and practices 6, 7, and 8 during the final discussion of the solutions they propose. In some

Box 2.2. Science and Engineering Practices in the _NGSS_

1. Asking Questions and Defining Problems

2. Developing and Using Models

3. Planning and Carrying Out Investigations

4. Analyzing and Interpreting Data

5. Using Mathematics and Computational Thinking

6. Constructing Explanations and Designing Solutions

7. Engaging in Argument From Evidence

8. Obtaining, Evaluating, and Communicating Information

problems, the research component may include practice 5 as they find and process data, practice 3 if they conduct hands-on experiments, and practice 2 if they use models to explore or explain certain phenomena.

While we acknowledge that the practices are woven throughout each of the problems, we have also identified specific key practices that are strongly emphasized in each content chapter. These key practices are listed in the tables describing the alignment with the _NGSS_.

Disciplinary Core Ideas and Performance Expectations

Another dimension of learning in the _NGSS_ is disciplinary core ideas (the third dimension in the _Framework_). This is a very important dimension for teachers to consider in their lesson planning, because the DCIs correspond to the content standards teachers must address. The _NGSS_ present these ideas as statements of scientific ideas and label them with a code that designates the content area, a DCI number, and the grade level.

As an example, one of the DCIs mentioned in Chapter 6, "Rock Cycle and Plate Tectonics," is ESS2.A. This is part of the second (2.A) Earth and Space Sciences (ESS) concept in the list of middle school standards (MS). The text of the DCI states the concept that students are expected to understand: "All Earth processes are the result of energy flowing and matter cycling within and among the planet's systems. This energy is derived from the Sun and Earth's hot interior. The energy that flows and matter that cycles produce chemical and physical changes in Earth's materials and living organisms."

The _NGSS_ also include performance expectations that are associated with the DCIs. The performance expectations are descriptions of indicators that students understand the DCI. These indicators are very helpful to teachers because they define the tasks and performances that can become both the activities and the assessments of learning we plan for the classroom. Using the same example from above, the performance expectation cited in Chapter 6 is *MS-ESS2-2:* "Construct an explanation based on evidence for how geoscience processes have changed Earth's surface at varying time and spatial scales." The performance expectation describes a behavior that reflects one of the practices (SEP 6: Constructing Explanations and Designing Solutions) relating directly to the DCI about using the geologic time scale to organize Earth's history. This same structure is reflected in all of the DCIs and associated performance expectations in the *NGSS*.

The performance expectation quoted above, along with the others cited in Chapter 6, were selected by the authors of the problems presented in Chapter 6 because teachers in the PBL Project for Teachers identified the concepts of matter cycling and the rock cycle (MDOE 2007) that they felt they could strengthen in their teaching. The state standards correspond well with the *NGSS*, even though the newer standards were written five years after the state standards. We expect that you will find that your state standards have a great deal of overlap with the *NGSS* as well.

Crosscutting Concepts

The *NGSS* identify seven crosscutting concepts (see Box 2.3)—ideas that span across multiple science and engineering disciplines and are useful as a way for students to connect their understandings in an integrated fashion. These CCs are more than important concepts; they can be guides for making sense of new material. For example, in Earth and space science, one of several important CCs is Systems and System Models. The movement of planets, the Sun, and the Moon represents a system that students should understand, and the interactions of tectonic plates at convergent, divergent, and transform boundaries also represent systems. Another system relevant to Earth and space science is the water cycle or hydrosphere. Thus, the CC of Systems and System Models helps students connect these phenomena that may seem unrelated at first glance. The CCs described in Appendix G of the *NGSS* are shown in Box 2.3.

To help you align your teaching with the *NGSS*, we have explicitly identified the CCs expressed in the problems we have developed. Chapters 5–8 each contain a set of related problems addressing science concepts from a specific content strand: Earth's landforms and water, rock cycle and plate tectonics, weather, or astronomy. As you consider your curriculum planning, look for the list of CCs at the beginning of each problem in the table presenting the alignment with the *NGSS*.

Accurate Understanding in a Self-Directed Process

One of the concerns we have heard expressed by teachers relates to students' exposure to inaccurate concepts during the initial discussions of the problem. The PBL framework asks students to make a "What do we know?" list, and sometimes learners will make statements

Box 2.3. Crosscutting Concepts in the *NGSS*

1. Patterns

2. Cause and Effect: Mechanism and Explanation

3. Scale, Proportion, and Quantity

4. Systems and System Models

5. Energy and Matter: Flows, Cycles, and Conservation

6. Structure and Function

7. Stability and Change

based on prior knowledge that might not match current understandings of science. For instance, participants involved in the astronomy problems made inaccurate claims about the cause of changes in the phases of the Moon. Some teachers fear that hearing these misconceptions might be contagious or could plant seeds of ideas that we want to avoid.

But in reality, learners already come to the science classroom with their own ideas and conceptual understandings, regardless of whether we discuss them in class. The PBL framework uses the "What do we know?" list as a way to organize ideas, and it can be used as an assessment tool to reveal prior ideas. Rather than avoiding discussion of prior ideas, PBL offers a process to examine those ideas and find out if they align with evidence and scientific theories. The process allows learners to challenge and modify their understanding rather than merely introducing new and competing ideas to the individual's schema for understanding the world.

In the process of facilitation, our enactment provides a way for the teacher to handle concepts that are not scientifically accurate. Facilitators are encouraged to not reject learners' comments during the analysis that accompanies Page 1 and Page 2 of the story. This is a challenge when students present ideas that the teacher recognizes as inaccurate. To deal with this situation, we implemented a practice of establishing some guidelines for the analytical discussion. One of these rules is that any idea listed under "What do we know?" needs to be verified or confirmed by information from the PBL story or from another reliable source. Ideas only supported by "I've always heard that …" are placed under the "What do we need to know?" or "Hypotheses" column. Later, as students search for sources or carry out investigations, the group can revisit the lists and cross out those they find are not supported by theories, laws, and evidence. Through this mode of learning, students learn *why* some ideas are inaccurate rather than merely being told their ideas are wrong.

Chapter 3 discusses this issue in more detail, including this manner of supporting claims about what we know for guiding the discussion. But teachers using PBL should not shy away from discussion of inaccurate assumptions. The process of working through "wrong ideas" is essential in bringing about conceptual change. Accepting that their ideas may be disproved is one of the attitudes we want students to develop. In this way, PBL helps teach and support many aspects of the nature of science (Lederman 1999).

References

Lederman, N. G. 1999. Teachers' understanding of the nature of science and classroom practice: Factors that facilitate or impede the relationship. *Journal of Research in Science Teaching* 36 (8): 916–929.

Michigan Department of Education (MDOE). 2007. *Grade level content expectations.* Lansing: MDOE. *www.michigan.gov/documents/mde/SSGLCE_218368_7.pdf* (for K–8); *www.michigan.gov/documents/mde/Essential_Science_204486_7.pdf* (for high school).

National Research Council (NRC). 1996. *National Science Education Standards.* Washington, DC: National Academies Press.

National Research Council (NRC). 2012. *A framework for K–12 science education: Practices, crosscutting concepts, and core ideas.* Washington, DC: National Academies Press. *www.nap.edu/catalog.php?record_id=13165#.*

NGSS Lead States. 2013. *Next Generation Science Standards: For states, by states.* Washington, DC: National Academies Press. *www.nextgenscience.org/next-generation-science-standards.*

FACILITATING PROBLEM-BASED LEARNING

The experience of being the teacher in a science classroom during a problem-based learning (PBL) activity is a bit different from what you might experience for other types of lessons. In some learning activities, your role is that of content expert or presenter of information. The students might be involved in recording information, listening, or perhaps applying new ideas. Alternatively, students might be carrying out some kind of science investigation as you direct and guide with questions. These roles are certainly appropriate, but PBL requires something different.

In PBL, the teacher definitely steps away from the lead role and instead becomes a *facilitator*. Educators use this term a lot in teaching, but for our model of PBL, we believe this role is accentuated. The facilitator's role is to provide minimal information but to provide resources and ask questions to guide the process. The students become more active participants in the discussion and even take the lead in identifying next steps and issues that need to be explored.

These new roles take practice—for both teacher and students. Students need to take risks in sharing and defending their ideas using information and evidence. Your role requires skillful questioning to guide without leading, and just as important, the ability to say nothing and let students explore their own ideas to find their misconceptions. In this chapter, we will use a vignette format to provide examples of what you might see in a classroom in which PBL is being taught, with a focus on how the teacher can guide discussions during the lesson. We will also share tips and strategies for successful facilitation of a PBL lesson; additional tips are provided in Chapter 4, "Using Problems in K–12 Classrooms." Some of what we share in this chapter is the result of our research on effective facilitation of PBL (Zhang et al. 2010), and some is based on our personal experience and teaching styles.

Remember, as you implement the lessons you select from this book, you may find that you need to practice your role as a PBL facilitator, and it takes time and practice to learn how to respond to students' ideas on the fly.

Moves to Make as You Go Along: Stage-Specific Advice

Facilitating PBL problems feels very different from traditional teaching and may require some strategies that are not part of your normal routine. Throughout this chapter, we will offer some "moves" you can plan to make. These are deliberate tactics to help your students think and talk about the problem they are analyzing, and the tips help you move into facilitator mode. It can be hard to remember that your role has shifted. You need to hold in some of your expertise and let your students struggle a bit with the challenges of solving a real problem. It is hard to do this, because you want to help them, but in the long run, stepping into the role of facilitator will help your students gain confidence and skills they need to think critically. And that's an important goal!

At the same time, there are times when the teacher needs to share his or her knowledge of the concept. This may mean giving some examples of phenomena that demonstrate a process or explaining how certain ideas are connected. The teacher also may need to ask questions to informally assess students' understanding or clarify what a student means by a comment or question. These moves are important in facilitating students' analysis of a PBL problem and in helping students make sense of the information they are finding. Part of the art of facilitation is learning when to use your content knowledge and when to hold back and let students explore an idea. For the beginning facilitator, we recommend patience: If in doubt, let students work for a bit, and then share your expertise.

Explaining Discussion Guidelines

Because you and your students may be experiencing PBL for the first time, it is important to set some guidelines for a PBL lesson. Discussion about real-world problems may reveal some strong opinions, some misconceptions, and some differences in beliefs and values that may be difficult for younger learners to understand. Before you start a PBL lesson, at least until your students learn to operate in this new type of lesson, setting some guidelines will help you manage the discussion and keep the conversation on task and respectful.

In the first section of the vignette, Ms. Sampson shows the class a list of guidelines for discussing PBL problems. These guidelines are useful in creating a climate in which participants are able to share ideas, pose questions, and propose hypotheses. They may also help to create a culture of open discussion in your classroom. Throughout the vignette in this chapter, we have tried to indicate how the science and engineering practices (SEPs) and the crosscutting concepts (CCs) from the *Next Generation Science Standards* (*NGSS*; NGSS Lead States 2013) appear in this lesson. See Chapter 2, "Alignment With Standards," for a complete list of the SEPs and CCs.

Ms. Sampson's Science Classroom: Discussion Guidelines

Ms. Sampson has been planning since the summer to try a new lesson idea. Today she's starting a PBL activity that she thinks will take about three days for her seventh-grade science class to complete. The topic is weather forecasting in her "Weather" unit, and today's activity follows some readings about weather and a video about the importance of weather forecasts.

Ms. Sampson: Class, today we're going to prepare you to make a weather forecast. As we work, you will take the role of a team of meteorologists, and you need to learn some information that will help you report the forecast on TV. We are going to use problem-based learning to look at this topic, so we need to set some discussion guidelines.

She projects a slide with the guidelines and discusses the list (see Box 3.1).

Box 3.1. Guidelines for Discussion

1. Open thinking is required—everyone contributes!

2. If you disagree, speak up! Silence is agreement.

3. Everyone speaks to the group—no side conversations.

4. There are no wrong ideas in a brainstorm—respect all ideas.

5. A scribe will record the group's thinking.

6. The facilitator/teacher will ask questions to clarify and keep the process going.

7. Support claims with evidence or a verifiable source.

Helping Students Function in a Self-Directed Classroom

This recap of discussion guidelines is important to help students start to manage their own learning. Although the PBL framework introduced in Chapter 1 is a good foundation for critical thinking, students may not have experience using a structured process for solving problems. In essence, we are making the metacognition needed to support learning more explicit (Bandura 1986; Dinsmore, Alexander, and Loughlin 2008) in a process that will help students develop the type of self-directed learning abilities we hope all our students can achieve.

The guidelines are important in helping students develop the habits of scientific discourse. A conversation in a scientific context is different from a conversation with friends about sports, music, politics, books, or other topics. So to help our students learn to function in a scientific community, or even just to understand the process behind scientific claims they might read about in an online news source, they need to know how we share and develop ideas in science.

At the same time, the guidelines are a reminder to the facilitator about his or her role in the discussion. As the facilitator, one of the most difficult tasks is avoiding the urge to give "right answers" to your students. But it is important for you to set an example by respecting new ideas or ideas you are uncertain about. Your role, especially at the beginning of a PBL problem, is to ask questions to clarify, to solicit responses from students who may be hesitant to share ideas, and to be the "referee" when the class rejects one student's ideas before any evidence has been discussed.

Recording Information

In the guidelines that Ms. Sampson shares, she mentions the "scribe." It is important to have a durable record of the ideas students generate. The written copy of the ideas students generate is also important as a "map" that students and the teacher can follow to see the development of their understanding. In a sense, posting the ideas as a list makes the learning "visible." The facilitator will use this list to make choices about guiding questions, information search strategies, and activities that can support the type of learning each particular class needs.

In some cases, you may wish to have a student serve as the scribe, but this may pull that student out of the conversation. It is difficult to create or share your own ideas when you're busy writing others' ideas on the board, and your students are probably

> **TECHNOLOGY TIP**
>
> SMART boards (interactive whiteboards) and similar technology are a good option for recording group discussions! They allow you to record a "page" of notes, move to a new page, and return to previous notes when needed.

not able to juggle those tasks. In our experience, it is best if you, the facilitator, can record students' statements, questions, and hypotheses on large sheets of paper, on the board or projected on the screen so all students can see the lists (see Figure 3.1).

You can create areas in your recording space for each of the three categories of ideas in the PBL framework ("What do we know?" "What do we need to know?" "Hypotheses"), but we suggest you use large pieces of paper taped to the board or the wall. This will let you add pages as the students' list of ideas grows. You can make notations or cross off statements and hypotheses as the students find new information, but it is important to have those items to look back at during the process of working through the problem. Students can see how their understanding develops, question why they think an idea is true, and connect the evidence with their new understandings. The large pieces of paper or electronic files will also allow you to move back and forth between different sections, if you teach the subject more than once per day.

Launching the Problem

Once you have established discussion guidelines and procedures, it is time to launch the problem. For this stage, you can have students arranged in small groups, seated on the floor in a circle, seated in desks, or whatever arrangement works best for you.

In Chapters 5–8, each PBL problem begins with an overview that describes the key concepts of the problem and aligns the problem with the three dimensions of the *NGSS* (NGSS Lead States 2013). This alignment includes a table describing the SEPs, disciplinary core ideas, and CCs addressed in the lesson. Keywords and a context for the problem are also offered to help you identify the problems that are most appropriate for your curriculum.

Figure 3.1. Recording Learners' Ideas in the PBL Framework

Following the overview and alignment page, each problem includes the text for The Story arranged in two parts. Page 1 is the part of the story you will use to launch the activity. Most of the stories are short and can be printed on a half sheet of paper. In some cases, you might project the story on the screen, but we find that it is helpful to give each student or group a hard copy so they can refer to it as they work through the analytical framework. You may choose to print one copy per student or let pairs or small groups read from the same page.

Start by handing out the copies of Page 1, and ask your students to read the story quietly. You might need to make accommodations for English-language learners or special needs students. Once everyone has had time to read through the story, ask one person to read the story aloud. This may seem redundant, but it is actually a very important step. Our research has shown that groups that read both silently and aloud at the start of the story generate a significantly higher number of ideas, questions, and hypotheses than groups that only read the story silently. We posit that in the first reading, students are working to comprehend the story, and in the second reading, they begin forming their own ideas in their minds. The time to process the story and think quietly seems to be important in supporting the discussion in the group as they move forward. The vignette sections that follow provide examples of how this process looks in the classroom setting.

Ms. Sampson's Science Classroom: The Launch

Ms. Sampson: OK, class, today's PBL is called Leave It to the Masses. Here is Page 1. Please read this story quietly. I'll give you about two minutes.

She hands out Page 1 of the Leave It to the Masses problem. (See Chapter 7, p. 174, to read the story.) As her class reads, she tapes three large pieces of paper to the board, labels them "What do we know?" "What do we need to know?" and "Hypotheses," and gets her colored markers ready. After two minutes, she asks for a volunteer to read the story. David volunteers, stands, and reads the story aloud.

Ms. Sampson: Thanks for volunteering, David. Now that you've heard the story, let's look at our three categories on the board. What do we know about the story right now?

The class is quiet for a minute, but she notices the students look like they are thinking.

Andrea: We are going to have to explain four different weather maps on TV.

Ms. Sampson writes Andrea's comment on the "What do we know?" paper.

Ms. Sampson: OK, good. What are the four maps?

Andrea: Temperature, wind, Doppler radar, and a map with highs and lows marked.

Jamal: What is Doppler radar? I hear that a lot, but is that different from other radar?

Ms. Sampson: OK, "Doppler radar" goes under "What do we need to know?"

Marcus: The story says air masses take on the characteristics of the surface if they stay there for very long.

Mai: And the map with highs and lows … I'm not sure what the highs and lows are showing.

Ms. Sampson: Mai, I can add that to the "What do we need to know?" page, too. Good question!

David: Well, down here it says something about high pressure. Warmer air is less dense, so it has lower pressure than cooler air.

Ms. Sampson adds David's comment to the "What do we know?" list.

Carmela: There's also a definition of air mass. It's a body of air with similar temperature and moisture properties. I'm not sure what "body of air" means … doesn't it all mix?

Ms. Sampson: I can put that under "what we know," but do you want to put something about "body of air" under "need to know"?

Carmela: Yeah, that's a good place for that.

Ms. Sampson: Great! OK, let's keep going. What else do we know?

The class continues to add more ideas to the pages on the board.

Moves to Make: "Unpacking Ideas"

During a discussion in the three-column framework described earlier, students are very likely to bring up terms and concepts that need to be "unpacked." *Unpacking* is a term commonly used in education and business conversations, but it is not always clear what unpacking an idea entails. In essence, students are using one of the SEPs as they analyze and interpret the information they are given (SEP 4: Analyzing and Interpreting Data). Students also use this stage to define the problem (SEP 1: Asking Questions and Defining Problems).

Let's focus on an example from the preceding vignette section. Mai brings up an idea to include in the "What do we need to know?" column:

> **Mai:** And the map with highs and lows ... I'm not sure what the highs and lows are showing.

> **Ms. Sampson:** Mai, I can add that to the "What do we need to know?" page, too. Good question!

The concept of "highs and lows" is certainly important to the problem about forecasting weather. But it is clear from Mai's question that not all the students in the class are familiar with it or know why it is relevant to weather forecasting. Ms. Sampson steers this comment to the "What do we need to know?" list and moves on.

It may be easy to imagine a discussion of highs and lows later in the lesson, but another useful strategy would be to "unpack" the concept right away. This can be done with questions that draw on what the students know about it already. These questions could be asked during the initial discussion, or they could wait until the class starts to explore the "What do we need to know?" list in more detail. But there are a couple of different ways to handle the discussion unpacking the concept.

Let's compare a "teacher as expert" approach with a "teacher as facilitator" approach (see Table 3.1, p. 24). In the "expert" role, the teacher shares what she knows, and the students become passive recipients. In the "facilitator" example, Ms. Sampson pulls information from the students, and the students' role shifts to either the expert or the problem solver who recognizes the need to find information. In the latter example, the students are active learners and consumers of ideas, a role we want students to master.

In the facilitator example, the students get much of the same information, but they have either discovered or remembered the information on their own and in their own words. The students have begun to develop some independence in learning and are practicing the skills used by proficient problem solvers. Independent learners can do more than just recall and repeat ideas. They synthesize ideas from information they are given or collect themselves (SEP 4: Analyzing and Interpreting Data). To demonstrate deep

Table 3.1. Comparison of "Teacher as Expert" Approach With "Teacher as Facilitator" Approach

TEACHER AS EXPERT	TEACHER AS FACILITATOR
Mai: And the map with highs and lows … I'm not sure what the highs and lows are showing. **Ms. Sampson:** The highs and lows are areas where there is either high or low barometric pressure. **Denise:** How can some areas have different pressures? Won't they all mix? **Ms. Sampson:** Well, each air mass has its own characteristics. Some have cool and dry air, and that means they have high pressure. Areas with warmer, wetter air have lower pressure. **Mai:** How do they end up so different? **Ms. Sampson:** That's because of the uneven heating of Earth's surface. Here, let's take a look at a diagram to explain that.	**Mai:** And the map with highs and lows … I'm not sure what the highs and lows are showing. **Ms. Sampson:** Mai, I can add that to the "What do we need to know?" page. Good question! Does anyone else have more information about that? **David:** Well, down here it says something about high pressure. Warmer air is less dense, so it has lower pressure than cooler air. **Denise:** And dry air is denser than wet air. **Ms. Sampson:** OK, so it sounds like we're talking about high and low pressure, right? What else do you know about high- and low-pressure air? **Steven:** Hey, when there is a low coming toward us, they usually say it's going to rain or storm. **Marcus:** Yeah, those are the Hs and Ls on the weather map. But I'm not sure how air can have different pressures in different areas. Or even wetness or temperature. Won't the air all mix? **Denise:** Well, there's this other part of the story that I think is important here. It says that if an air mass stays in an area for long, it takes on the characteristics of the surface below. So I'm guessing if air is over a desert, it gets hot and dry. If it's over a big lake, it gets wetter. And it sounds like that tells what the pressure will be. **Mai:** Oh, so is that why we always get more rain and snow next to Lake Michigan? Does the air get wetter there?

understanding, students should be able to synthesize information by connecting ideas in the context of a real problem instead of repeating bits of disconnected facts. In the expert example, Ms. Sampson is hinting toward the concept of high- and low-pressure air masses, but we cannot tell if students are building their own understanding of the concept and the problem.

Generating Hypotheses

As students work through the analytical discussion of Page 1, they are likely to state ideas that reach beyond "What do we know?" and "What do we need to know?" In the next section of the vignette, watch for the comment that suggests an inference. Sometimes these are subtle, but as the facilitator, you can point out the step the student has made and suggest adding this new idea to the list of "Hypotheses."

As a facilitator, you will need to pay attention to the questions students ask during the discussion. One common pattern is that learners will present an idea as a question when they have some uncertainty about the statement. A student may suggest a question to add to the "What do we need to know?" list, but the question is actually a tentatively worded hypothesis. Let's look at an example of this.

Ms. Sampson's Science Classroom: Generating Hypotheses

Ms. Sampson: OK, class, you've cleared up some ideas about air masses, so let's keep working. Any other things we need to learn about or ideas about this problem we should add?

Angie: I have a "need to know" thing. Can we use temperature and air pressure to predict when a storm is going to happen? I think we can, but I'm not sure.

Ms. Sampson: Good question, Angie, but I think I hear a hypothesis in that statement. You're asking if temperature and pressure are good predictors, but if we reword that, can we make this a hypothesis?

Angie: I'm not sure if I'm right, though. I'm not sure this is a good hypothesis.

Ms. Sampson: But that's OK, Angie! Remember, a hypothesis is a proposed answer to a question that can be tested, and if the evidence eventually shows that it's not correct, that's alright! So do you want to try to build a hypothesis from your question?

Angie: I guess so. I'm not sure how to start it, though. "I predict that temperature and air pressure are the ..." Is that the way to state it?

Carlos: Shouldn't we use the same kinds of words we use in other labs? *"If, then,* and *because?"*

Ms. Sampson: That's what we use when we're going to change a variable and see what the result is, Carlos, but that's a start. Who remembers what we use when we're observing events instead of changing a variable?

Alyssa: Isn't that when we use the "I think that …" kind of hypothesis?

Ms. Sampson: Yes, that's right, Alyssa! So, Angie, use that as a start. "I think that …"

Angie: OK. "I think that temperature and air pressure are the best data for predicting when a storm is going to happen."

Joseph: It needs a "because" statement.

Ms. Sampson: Yeah, what would be the "because" part?

Angie: Because it usually gets colder before a storm, and the weatherman said the air pressure drops before a storm.

Ms. Sampson: Good! That's our first hypothesis. Can you tell me more about what's going on with temperatures?

Andrea: OK, so a warm air mass moving over cold air creates a thunderstorm?

David: No, that's not what happens. I don't think that's right.

Ms. Sampson: Remember, we're making hypotheses. We need evidence before we can reject a hypothesis, so I think we need to include it on the "Hypotheses" page.

Carlos: I have a different hypothesis. I believe the best way to predict a storm is to look at wind speeds and air pressure. And colder air is denser than warm air, so I don't think cooler air means a low-pressure area is coming.

Ms. Sampson: You need to put it in hypothesis form, too!

Carlos: How about this? "I believe wind speed and air pressure are the best ways to predict a storm, because the air pressure drops and the winds pick up right before a storm."

In this example, a student initiated the first hypothesis, but it began as a "What do we need to know?" question. Note the way that Ms. Sampson directed the discussion toward the "Hypotheses" column in the analytical discussion and pointed out that Angie's question seemed to include a hypothesis. This is a very common pattern in the discussion of Page 1 with most problems, and you need to watch and listen for those types of questions. One cue is to look for a "because" statement in the question. For instance, if a student says, "I want to know if wind speed predicts storms, because it always seems to get windy just before it storms," this suggests a hypothesis. The "because" indicates a connection between cause and effect (CC 2: Cause and Effect: Mechanism and Explanation) or a rationale for a possible solution to the problem. (SEP 6: Constructing Explanations and Designing Solutions). The teacher could easily leave the question worded as it is, but it helps to move it to the "Hypotheses" column. Students can then "test" the hypothesis as they do information searches later in the lesson.

The strategy Ms. Sampson used was to point out the purpose of a hypothesis and mention that the question asked sounded like a testable question. She then asked students to rephrase the question rather than doing the rephrasing herself. This puts more control over the process in the hands of the students so they must practice this skill. Ms. Sampson is truly taking the role of facilitator by steering students with questions and letting the students generate the final version of the hypothesis. This facilitating includes reassuring Angie that it was okay to hypothesize and later find that the hypothesis is not supported. You've probably seen students' reluctance to be "wrong" on a hypothesis, and PBL helps them get over that fear.

It helped that Ms. Sampson's class had learned a deliberate pattern for writing hypotheses in other lessons. If you have been working on SEP 3 (Planning and Carrying Out Investigations), your students will likely have begun learning this skill as well. In your class, part of the scaffolding you will do with students is to help them learn to ask questions, write hypotheses, build data tables, and write explanations. PBL gives you yet another context in which students can use those same practices, so you have the flexibility to insert your particular format for structuring these elements of the science process.

Angie's hypothesis took quite a bit of scaffolding. Students contributed bits and pieces and made connections with the class "standard" for hypothesis writing. It was not an automatic process at first. This is typical of students who are still learning to think like scientists. Carlos was able to phrase his hypothesis in the appropriate format much more quickly because he was part of the process of working out that format during the discussion about Angie's hypothesis. This is also a common event. Students very quickly adopt the structure when the class works through the process out loud and can see the hypothesis on the list as a reference for later discussion.

If no students come up with hypotheses on their own, the teacher needs to help students think about making some predictions or proposed solutions. As the list of "What do we know?" and "What do we need to know?" items grows, a facilitator can ask something like, "So what do you think is the answer to the challenge at this point?" This is usually enough to get the ball rolling with the first hypothesis.

Our experience suggests that once the first hypothesis emerges, other students become more comfortable suggesting possible solutions or hypotheses. In other cases, students may need a prompt from the facilitator. You can elicit hypotheses by asking, "So what do you *think* is the answer to the challenge?" or "Do you have any hypotheses about a solution?" If students are really having trouble framing an initial hypothesis, you can ask if they think there is a relationship between any of the things listed under "What do we know?" Defining relationships is often the beginning of a hypothesis. Such initial hypotheses may not be complete answers to the challenge, but they start the ball rolling.

Introducing Page 2

As your students work through the PBL analytical framework and the information on Page 1, there will be a moment when the students start to run out of new ideas to put in the three categories of the framework. They will exhaust the "What do we know?" ideas and address most of the learning issues on the "What do we need to know?" page. The list of hypotheses might be short, but the generation of these ideas will slow down. *When that happens, your job as the facilitator is to transition into Page 2.*

Page 2 continues the Page 1 story and adds new information that will help students work toward a solution to the challenge statement at the end of Page 1. Introducing Page 2 should work very much the way introducing Page 1 did; students will read Page 2 quietly, then a student will read it aloud. Once that happens, the class can repeat the analytical process, adding new ideas to the same three categories of the PBL framework.

One major difference in the way to handle information relates to the new content on Page 2. You may find that "What do we need to know?" items on your list will be answered with the Page 2 story, or that the hypotheses generated in the first discussion will be rejected based on the new information. You can certainly add new questions and hypotheses as well as "What do we know?" statements, but we strongly recommend that you keep the first set of ideas on the board and visible to students. As you answer items in the "need to know" list, cross them out but leave them on the list. Some facilitators keep a list of "summarized knowledge" under each question to connect the "need to know" items with the new information they use to answer the questions. When you learn enough to eliminate a hypothesis, don't delete or erase it, but cross it out. Having those ideas visible is helpful when students look at the path they have taken from their initial ideas to the final solution for the problem. Processing their own ideas this way gives students a way to know *why* the solution works, not just that this is the right answer. It also builds a habit for students to show their thinking and their work. You might even find that when students begin to adopt the PBL skills as habits, they apply them in other subjects as well!

Ms. Sampson's Science Classroom: Introducing Page 2

Jason: So we know there are different kinds of air masses, but I still don't know which of these are related to storms or severe weather or whatever they call it.

Ms. Sampson: So do you want to put that under "What do we need to know?"

Jason: Yeah, I think so.

Ms. Sampson: OK, got it. What else can we add to our lists?

(long pause)

Andrea: I think we need to find out if storms all have the same wind direction.

Jamal: We already have something about wind direction under "What do we need to know?"

Ms. Sampson: Yeah, I think we have that covered. Any other ideas? Or new hypotheses?

(long pause)

Ms. Sampson: OK, then it sounds like you're ready for more information, right?

(Multiple students): Yeah! We need more information.

Ms. Sampson: Alright then, here's Page 2. Let's do what we did with Page 1. Read the story to yourself, and then we'll read it out loud.

She hands out Page 2, the class reads it quietly, and Devin reads Page 2 aloud.

Ms. Sampson: OK, good. Now let's add new pages for "What do we know?" "What do we need to know?" and "Hypotheses." We need to talk about each of these pages again with the new information we have. So … what do we know NOW?

Will: Well, we have five days of weather maps that can show how the highs and lows move.

Rose: And there are lines on the maps with arrows and half circles on them.

Marcus: Wait, I'm not sure what those are. I think we need to put those lines under "need to know."

David: Those are fronts. My dad told me about those once when I asked him about it when we were watching the weather. There are hot and cold fronts, and that's what the lines show.

Marcus: Yeah, but which one is which?

David: Umm … I can't remember. Ms. Sampson, which one has the arrows on the line?

Ms. Sampson: Good question! Let's put that under "need to know."

Tricia: Did you notice that wherever it rains, there's a front with the half circles? And there's an *H* on one side and an *L* on the other.

Jason: Does that relate to the hypotheses we wrote? Carlos said he thinks storms can be predicted by looking at pressure and wind.

Carlos: Yeah. That line shows where a high and a low meet. I think that's where storms happen.

Denise: We need to check that pattern. Put that under "What do we need to know?"

Vince: Yeah, but I think we can even look at some data about barometric pressure. We could build one in class and record data. I am pretty sure the air pressure will match temperatures. They both either go up or down together. I saw something the other day where it shows how to make a barometer to measure the air pressure. You get a U-shaped tube and put some liquid in it. And one end has to be open.

David: Cool! Can we build one? Maybe I can Google this and get the instructions. We can find some tubes in my grandpa's garage. He's got all kinds of junk we can use.

Ms. Sampson struggles to let the conversation work its course—they are getting off track and starting to talk about issues that are not important to the problem.

Alyssa: So maybe we just need to look at the map that shows winds. That's one of the maps we have to talk about. The weather map always has those arrows for the wind direction.

Mai: Yeah, that's the second part of Carlos's hypothesis. Is there a pattern that matches the weather?

Ms. Sampson: That's a good idea. If you think it's important, do you want to make that a hypothesis? Vince, we can add your hypothesis about the pattern with temperature and air pressure if you like.

Mai: Yeah! I believe air pressure and wind speeds tell the weather guy when air masses are going to … what's the word I want? Meet? It's like they crash into each other.

Sarah: Yeah, and on the maps, you can kind of see where the air masses are. And the fronts are usually on the edges.

Ms. Sampson: Let's put those under "need to know," Sarah. OK, I have that info recorded. Do we know any other new information?

Carmela: Yeah. The diagrams with the maps say that thunderstorms form when cold dry air pushes under warm, moist air and that the warm air rising is why a storm happens.

Angie: Wait, does that mean my hypothesis is wrong? I was thinking that cold air and low pressure go together.

Devin: Does that mean we cross out that hypothesis?

Ms. Sampson: We could, but can we just modify it with this new information?

Angie: Yeah, just switch what I said about warm and cold air, and we can keep it.

Moves to Make: What If Students "Go Down the Wrong Path"?

In this section of the vignette, we see Ms. Sampson guiding the class through the analysis phase of Page 2. Students listed the new ideas they got from Page 2, raising questions about ideas they didn't understand and offering new hypotheses. But we also see an example of students "going down the wrong path." Some conversations take off on tangents, like the comments about making a barometer, and others may follow incorrect hypotheses that the teacher knows are going to lead to a dead end.

As the teacher, you will encounter those moments when you want to comment to prevent the class from following a "wrong" hypothesis. You should already know what some viable solutions to the problem are, and you simply want to help your students find the right answers. But it is important *not* to interject comments that stop students' exploration of incorrect ideas. A hypothesis that is later rejected is a powerful learning experience and is likely to lead to enduring understandings. So you need to let students explore those ideas, even when your instincts tell you to steer them in a new direction. Teachers are likely to want to correct the inaccurate ideas right away, but the PBL framework emphasizes letting students find evidence that leads them to eliminate ideas on their own.

Note how Ms. Sampson handled it. She allowed the class to work through their ideas, and she included Vince's hypothesis in the list. You should avoid eliminating hypotheses for your class. Let students decide when an idea is rejected. That's a difficult thing for teachers to do, and it may take some practice, but it is important! In this case, Vince helped the process by introducing a new hypothesis to compete with Carlos's hypothesis. Including them both will allow students to compare them using evidence and information they collect. Eventually the students will have all the tools they need to decide which is the most viable hypothesis.

When you encounter this type of situation, be assured that it's normal in the PBL process. Each of the authors has experienced this, and we have felt the same internal conflict between providing content knowledge or letting students learn or discover for themselves. We've all learned to be patient, let the students drive the discussion, and wait for the learners to see all the information before we simply give answers.

But there are good strategies for redirecting the discussion! One suggestion is to establish a practice in which you, the teacher, are free to participate as a learner. This gives you permission to ask the same types of questions students should be asking. In this co-learner role, you can model critical thinking and questioning while using your comments to keep students on task and on track.

Here are some questions or statements, or "steering tools," you can use to keep your class discussion on track:

- "So how does that apply to the challenge for this problem?"

- "Maybe we should restate the question we are trying to answer."

- "Do we have a source that can verify that idea?"

- "What kind of evidence do we need to support that?"

- "How does this information from Page 2 relate to Page 1?"

- "That sounds like a 'need to know' issue."

Researching and Investigating

Once your students have completed the discussion of Page 1 and Page 2, you should have an extensive list of items under the three categories in the PBL framework: "What do we know?" "What do we need to know?" and "Hypotheses." On some of the lists, you may have crossed out questions you've answered or hypotheses you've ruled out. The information that is left should point to learning issues and predictions that have potential as solutions to the challenge presented on Page 1. Remember, the goal is to propose solutions to the challenge, so the research and investigation should focus on this goal.

The next step in the process of facilitation is to help the class develop a plan for gathering information or conducting an investigation that will answer the "What do we need to know?" questions that are still unresolved. In this phase of the PBL process, the teacher has some choices that will determine what the next part of the lesson will include. Is there an inquiry-based lab or hands-on investigation that would help students understand the concepts that underlie the problem? Will students use a computer lab or classroom computers to search for information on the internet? Are there text resources that can help them answer the questions? Should the teacher provide a limited set of readings to ensure that students find productive information? All of these may be appropriate choices!

Investigations

In some problems, there may be a hands-on activity, such as a model that students can build, that would help illustrate a concept. For instance, in the "Weather" chapter (Chapter 7), the Northern Lights problem is an ideal situation in which to use a simple sundial or to have students build a Solar Motion Demonstrator to track the movement of the Sun through the sky in any given month. This allows students to experience a real-world phenomenon and use data as one type of evidence in constructing their final solutions.

You may also have inquiry-based investigations your students can conduct to learn or reinforce specific concepts. Astronomy problems like E.T. the Extra-Terrestrial (see Chapter 8) may be an opportunity to insert your favorite demonstration, model, or simulation of Moon phases. Problems from Chapter 5, "Earth's Landforms and Water," provide a context for doing a lab on water infiltration through different types of soil.

One of your roles as the teacher is to plan for these investigations. You may have activities in your textbook resources that would be appropriate, or you may find or create new lab activities to meet your needs. In Chapters 5–8, we have provided some lab activities that fit with specific concepts, including instructions to help you plan and implement these activities.

An important component of any activity is safety. Students and teachers need to learn how to properly assess risks and take actions to minimize risks. Safety issues to be considered include the use of sharp objects, use and disposal of chemicals, and the presence of fire or burn hazards. Teachers are responsible for precautions such as wearing safety goggles or glasses, providing disposal containers for sharps and chemicals, and ensuring that students know where fire extinguishers and chemical showers are located.

Information Searches

Some problems are best addressed by helping students search relevant resources for answers to the learning issues they have identified. For teachers who need to integrate literacy standards into science teaching, the skills of finding and evaluating information from multiple sources are clearly featured in this part of the PBL process.

Sources for answering the learning issues your students have identified may include web searches, their science texts, books in the school's library, or magazines and newspapers. Although our first thoughts seem to turn toward technology as the go-to source, there are many text-based tools that are certainly appropriate. You can decide which are best suited for the context in which you are teaching based on access, convenience, or the "fit" for the topic at hand.

The search for information also offers multiple choices for scheduling. Perhaps you will have students work on this the same day they have analyzed Page 1 and Page 2, or you may need to plan this phase for the next day or as homework. The number of days you spend on this task also depends on your specific needs.

Ms. Sampson's Science Classroom: Beginning the Information Search

Ms. Sampson: Alright, class, you've created a good list of facts, hypotheses, and things we "need to know." Now we need to plan what information we'll look for next. Let's look at the "What do we need to know?" list. Are there specific ideas that groups will offer to find out more about?

The students talk softly with their groups about what they want to research.

Jamal: Our group wants to know about that Doppler radar question. Can we look for that?

Ms. Sampson puts Jamal's name next to "Doppler radar."

Ms. Sampson: You got it, Jamal! Your group can get started.

Denise: We'll look for information on fronts and air masses.

Ms. Sampson: OK, Denise, that's a good topic to look at.

Jason: What about the air pressure idea? Is that part of one of those others?

Ms. Sampson: I don't think so. Do you three want to look that up?

Jason: Yeah, we'll take that topic.

Rose: What about wind speeds? We need to figure out how the wind arrows on the map match fronts and air masses. We can search for that.

Ms. Sampson: Good idea! If you'll volunteer, you can do that. Alright then, folks! You need to get started with the time we have left today, and we'll continue working on this tomorrow.

Angie: Can we look stuff up at home tonight, too?

Ms. Sampson: Sure! But make sure you write down what sources you find and bring the list with you tomorrow. Remember, when we're done, each group is responsible for sharing what you find with the entire class. Be organized!

Teacher-Selected Sources

For some classes, "searching" for information may require more assistance from the teacher. In these cases, the teacher might pick a limited collection of resources and provide these resources to groups when they are ready to find answers to their learning issues. Perhaps the problem is complex enough that you want to steer students to specific sources. Maybe the information they need is not easily accessible to your students, either because very little is published online about the topic or because your school filters access to the necessary sites. Even the age or technology skills of your students may suggest that you should preselect the sources.

One strategy for doing this is to create sets of articles or websites that address specific topics. You can either give each group of students all of the sets or distribute each set to different groups. The latter option forces students to read and analyze the texts and share what they find with other groups. This type of communication is common among practicing scientists and addresses skills that students need to develop across the curriculum.

To help you select problems for which preselected sets of sources are useful, we strongly recommend that you work through each problem in advance. Think of the types of "need to know" issues you expect students to identify, and try searching for those concepts. If you can't find them easily, your students may also struggle to locate sources. Many of the problems in Chapters 5–8 include a Resources page (Page 3) with links to websites and references to other materials that are relevant to the science concepts.

Sharing and Resolving the Problem

When your students have completed the investigation or information search, the next phase includes sharing what they found. If each group has selected specific learning issues to research, this sharing is critical to the challenge presented to the class. No one group is likely to find all the information they need to solve the problem or build a complete solution to the challenge. But if they share information, the class can co-construct some solutions, much as project teams do in the workplace. This phase of the PBL process gives students a chance to hone their skills with SEPs 6–8: Constructing Explanations (for science) and Designing Solutions (for engineering); Engaging in Argument From Evidence; and Obtaining, Evaluating, and Communicating Information.

The class sharing session should still focus on the three pages of analysis the students created during the discussion of Page 1 and Page 2, especially the "What do we need to know?" and "Hypotheses" lists. The information search should address specific "need to know" items, and their findings should help in the evaluation and adjustment of some of the hypotheses as students apply what they have learned to the challenge presented in the story. Post the three lists on the board or on a wall for all to see and take a minute to recap what the class has done so far.

Each group should be asked to share. Although some students may be reluctant to speak in front of the class, building their comfort with such a task is an important learning goal. We find that when the presentation is informal, the task is less threatening. One way to promote sharing is to ask a student in each group to share one thing they learned. This leaves room for others in the group to share their ideas. Sharing their findings also helps students learn to pay attention to evidence and reliable sources.

As groups present what they found, it may also help to have other students take notes or record concepts in a journal or science notebook. They should also be encouraged to ask questions that help clarify ideas. Let your class know that the goal is not to stump or quiz each other, but to help the entire class understand the information.

If your class or specific groups did an investigation, this is a good time to have the class look at the procedures and results and talk about what the evidence means. If you have a standard procedure for presenting scientific explanations from an investigation, this is a perfect time to apply that structure. For instance, you can establish a procedure in which students share observations and data, identify patterns in the data, and suggest an explanation for the patterns. In the case of developing a solution for a problem, another approach is to describe the proposed solution, explain why it will work, and explain how evidence supports the ideas. If you have a structure you use for this in your current lab activities, you can use the same structure with your PBL lessons.

When all the information has been presented, you have options on how to construct solutions. One way to come to a final answer to the problem or challenge is to discuss the problem as a group. The focus on this should be the hypotheses created by the class. When a group wants to support a specific hypothesis, you can ask for a rationale: What evidence makes you think this is a good hypothesis? Other students should also be allowed to make counterclaims about a hypothesis or to present ideas that would refute the hypothesis or solution. This discussion can be a rich assessment of students' learning and ideas because it forces students to reveal the connections they make between concepts as they apply them to an authentic problem. Recording their ideas may be helpful if you wish to assess these connections, or you may choose to have a checklist so you can keep track of evidence of new learning.

In some of the classrooms in which we have observed teachers using PBL lessons, we have also seen another approach. Some teachers elect to have each group talk about the evidence they have found and create their own solution to the problem. This works best if each group was responsible for looking up more than one concept from the "What do we need to know?" list. It is helpful to set a time limit for this discussion, and you may want to have a structure for the group's response, as described earlier in this section. The teacher may also have a handout with general questions for the group to answer. This can include what hypothesis the group was investigating, what "need to know" issue they explored, what evidence they collected through research or experimentation, and how the evidence leads to a solution. The group then presents their ideas to the class, and other groups are encouraged to ask questions or explain what they see as problems in the solution.

In both of these scenarios, the next step is to ask for a solution to the challenge listed at the end of Page 1. This is the ultimate goal of the activity, so make sure you pay attention to the challenge. Students might present more than one solution. That's okay! In the real world, there may be multiple ways to solve a problem, and we want students to understand that. But when more than one solution is presented, you can ask the class to discuss the strengths and weaknesses of each solution, ask them to vote on the one they prefer, or ask each student to write a short response or exit ticket with a prompt something like the following: "Which solution do you think is the most useful? Explain why you chose this solution over the others." (See the "Assessing Learning" and "Responses to Assessment Data" sections later in this chapter for more information on exit tickets.)

Ms. Sampson's Science Classroom: Sharing and Building Solutions

Ms. Sampson: Today we're going to share the information you found about the Leave It to the Masses problem we've been working on. As you present, remember that you need to describe the answers you found clearly, and you should be ready to tell us where you found them. We'll use that information to see what we can cross out on the "need to know" list and how your information fits with our "Hypotheses" list. I need each group to share what they found. Jamal, I'd like your group to start, if you don't mind.

Jamal: OK. We looked up how Doppler radar works. First of all, we found out that Doppler radar towers are the things that look like giant golf balls, like that one over by the old schoolhouse on State Road 13. I always wondered what that was. And it works like regular radar because it bounces radar waves off of objects to find out how far away they are. But the difference is that this kind of radar measures how fast and what direction an object is moving. If an object is moving away, the waves coming back from it get farther apart. If the waves are getting closer together, it's moving toward you. They call that the Doppler effect. So it can tell if clouds or rain or whatever is moving. It can also tell the difference between clouds, light rain or heavy rain, or hail. It can tell how hard it's raining by the size of the drops, I guess.

David: I thought the Doppler effect was what makes a train whistle change pitch when it goes past.

Ms. Sampson: It's the same idea, David. Radar uses radio waves, and the train whistle is sound waves. But in both cases, the distance between waves changes. In the train whistle example, you hear that as a change in pitch. OK, Jamal, that's a good start!

Denise, Angie, and Mai share information their groups found about fronts and air masses, including information about the symbols on the weather map for cold and warm fronts, and how they form when two different air masses meet.

Ms. Sampson: Good information, girls! We can use that as we look at the weather maps.

Jason: But how do we know whether a cold or warm front forms when the masses of air have different temperatures?

Carlos: Yeah, does the cold meet the warm, or the other way around?

Ms. Sampson: Good question. Let's look at the diagrams of cold and warm fronts on Page 2 of the story again.

Andrea: Can we put them up on the screen? I think we can see the answer in them.

Ms. Sampson: Sure, use the document camera.

Andrea: Look at the arrows on the bottom of the page. If the cold air moves toward a warm front, it's a cold front, and the warm, moist air rises and forms big storm clouds. If the warm air moves toward a cold air mass, it's a warm front, and the clouds are different.

David: You mean it's the direction the masses are moving?

Ms. Sampson: That's what our information shows, so yeah, it's the direction the air masses are moving. And if that's the case, can we find out what direction air masses are moving so we can use that for a prediction?

Jason: Well, yeah, we can. The wind map has arrows and lines that show what direction the wind is moving, so that's what we should use, right?

Rose: Are you sure? I thought our weather always moves from west to east, so I'm not sure wind direction causes that.

Ms. Sampson: Let's look at the data. Can you see air masses moving if you think of each day's map as part of an animated movie?

Marcus: Try putting two or three days of maps on the document camera and let's look.

Ms. Sampson: OK. Here are three days. Are there air masses that look like they are moving?

Mai: Kind of. I think they all do.

Ms. Sampson: OK, now let's look at the wind maps with these same days.

Angie: Oh yeah! The arrows match the movement of the masses.

Rose: Hmm ... I guess it does. And there's a high-pressure mass moving from south to north, sort of. That makes it pretty easy to see how air masses move. But they don't always move just west to east.

Ms. Sampson: OK, so I think we have a better understanding. Let's look at the rest of the information you found.

The class looks at a few more pieces of information and agrees that using the maps of temperatures and air pressure helps identify air masses, and the wind maps help predict movement of the air masses to determine how the interactions may cause changes in the weather.

Moves to Make: Correcting Misconceptions or Nonscientific Solutions

When your students are constructing and selecting solutions, they are considering information their class has shared, but they also are influenced by prior knowledge. Sometimes this prior knowledge is not accurate, and it is likely to be durable and difficult to change. These ideas can lead to solutions at the end of the analysis process that are not practical, fail to really solve the problem, create other problems, or omit concepts the teacher has identified as important learning goals.

So what should you do when that happens? Our first suggestion is to assume the role of a classmate by asking questions you know will force the class to think about an important concept or piece of evidence. When skillfully used, these kinds of questions can help students notice the problems with their claims. One of the most effective approaches is to have students compare a problematic claim to information they have listed under the "What do we know?" column of the analysis charts.

One of the strategies that can be very effective is to ask questions such as "Are there any 'what we know' statements that contradict this solution?" In the vignette, Ms. Sampson asked students to review the weather maps over three days to look for patterns in the movement. By asking students to compare their researched information with other facts and evidence, you can help them develop SEP 8: Obtaining, Evaluating, and Communicating Information. This is a critical practice in our world of abundant information. Students will be exposed to many claims and proposals in the news, at work, through advertising, and in legislative bills that need critical analysis against the available evidence. This also helps address at least two of the "Essential Features of Classroom Inquiry" listed in the National Research Council supplement to the *National Science Education Standards* (National Research Council 1996, 2000) by asking students to give priority to evidence as they form and evaluate explanations.

Another approach would be to ask students to list the strengths and weaknesses of each solution. As in the strategy above, this places students in the role of evaluators and requires comparison of solutions to evidence. This also models the type of analysis used in the workplace for problems related to science and engineering, as well as many other contexts. Remember, the phase of the PBL process in which students generate solutions highlights both synthesis and critical thinking, so having students engage in these types of thinking is important.

But what if this doesn't do away with a misconception? Or what if the class didn't grasp a key concept that makes a big difference in the problem? Scientifically incorrect ideas can be durable and may get in the way of students' assimilation of new ideas. Some of the peripheral information may draw students' attention as they create solutions. So the teacher needs to be prepared to correct ideas and guide the development of solutions during this final part of the PBL lesson.

When your students just aren't applying concepts accurately, you now have a chance to explain ideas. There are times when your students need you to be the expert. Although we suggest you be patient with students' own thinking process, you may need to step in and present information that students need. If needed, you can lecture, lead a discussion, show a simulation or an image, or introduce some type of activity to help guide the learning. A good example of this is illustrated in the vignette when Ms. Sampson explained the connection between Doppler radar and a familiar example of the Doppler effect. This phenomenon is a key concept in understanding how Doppler radar is used to view weather data. Direct teaching has its place in the classroom, and your content expertise is important. If Ms. Sampson felt her students needed more information about the Doppler effect, she could use this opportunity to present a short lecture or demonstration about waves, wavelength, and frequency, with examples relating to light, sound, and radio waves.

Assessing Learning

When implementing a PBL lesson, the teacher/facilitator should respond to the learning needs of his or her students as they emerge. Flexibility is key, but to be flexible the teacher needs information about what students are thinking. Assessment is an important part of the facilitation process. As you lead a class through PBL problems, you should be planning to assess and to use the information from your assessments to adjust your teaching.

The PBL process as we have described provides for continuous assessment. The process of analysis using the PBL framework allows the teacher to hear and see what students are thinking as they talk about their ideas and record information, questions, and hypotheses under the three columns of the analytical structure. Each comment from a student gives you insight into their understanding.

But be aware that what you hear in a group discussion may not reveal what every individual is thinking. In a whole-class discussion, the teacher sees a "group think" picture of what students know. There may be bits of information from a handful of students that seem to make sense when the entire group shares ideas, but you need to know what each student understands. It is helpful to have strategies that let you assess individual students rather than the entire group of students.

The need for individual assessments is even more pronounced if the activity takes more than one class period. As we developed our model in the PBL Project for Teachers, our facilitators found it very helpful to implement informal assessment strategies like exit tickets. These are very brief prompts asked before the end of a class period for which students write a short response. These prompts may focus on one idea the students learned, one idea they found confusing, or one question they have based on what happened in class. You might also ask students or groups to give a written summary of the information they found during their research, their choice of the "best hypothesis so far," or a drawing of the concept they are exploring.

Another form of assessment is the transfer task. *Transfer of knowledge* refers to the ability of students to apply knowledge of the concept in new contexts. For instance, students may know that air masses take on the characteristics of the surface where they develop, but they should also recognize that the extreme low pressure and high winds of a hurricane are more likely to form when an air mass is positioned above very warm ocean water where the moisture can evaporate and be contained in a warming air mass. The importance of transferring knowledge to new situations is supported by Schwartz, Chase, and Bransford (2012), who suggested that a deep understanding of a concept must be accompanied by transfer. To help you perform this type of assessment, the problems in Chapters 5–8 include transfer tasks. The transfer tasks are often used as a summative assessment, but they can also inform the choices the teacher makes about the next activities to include in a unit.

In Chapters 5–8, we also present open-response questions that we have developed and tested for each content strand. There are two types of these questions—general and application—to address the concepts and standards included for the problems in the content strand. We discuss the role of these assessments further in Chapter 4, as well as options for when to use the assessments and how to interpret responses.

Responding to Assessment Data

Assessment of learning is important, but you also need to consider how you can use the assessments to respond to students' needs. We've introduced a couple of assessment strategies that can help you select your next moves as a facilitator in the PBL lesson. But it may help to share some examples. These examples include exit tickets and group summaries of solutions to PBL problems.

Exit Tickets

This is a simple and quick way to collect information about your students' understanding and issues that need to be resolved. Exit tickets (Cornelius 2013) can ask one of several different kinds of questions, including "What's one thing you've learned?" "What about today's topic are you still confused about?" or "What's one question you have about today's lesson?" Each student then writes a short response and turns it in to the teacher at the end of class. The next step is for the teacher to read through the tickets to see if there are important issues that need to be handled in the next day's class.

The following vignette section provides an example of how this might work in Ms. Sampson's class.

Ms. Sampson's Science Classroom: Exit Tickets

Ms. Sampson asked her class to write exit tickets after Page 2, using the prompt "one question you have about the Leave It to the Masses problem."

Ms. Sampson: OK, I looked over the exit tickets you wrote yesterday, and I think we need to add something to the "need to know" list. Several of you wrote that you don't understand why dry air is denser than moist air. Can we add that to our list?

The class agrees, so this is added to a list of topics to be researched.
Another possible result might be …

Ms. Sampson: Your exit tickets tell me that there may be some questions about the connection between air pressure and the direction the winds move around a low-pressure area.

She explains that air is a fluid that moves like water and that the rotation of Earth helps create forces that push air around. As air moves, it is pushed toward a low-pressure zone by the higher pressure in other areas. She uses a 2-liter bottle of water to show how water moves more quickly if it swirls in a vortex as it drains out of the bottle when it is turned upside down. She relates the movement of air toward the low-pressure zone to the movement of the water, to help students visualize the concept.

Group Summaries

In the PBL Project for Teachers, we found that an entire class may agree to a solution, but some individuals may have a different level of understanding of the concept. One of the strategies we tested proved to be useful—group summaries.

In this assessment, each group is asked to write a summary of their group's proposed solution. The summary should include a description of the solution they think best solves the problem or answers the challenge, along with a rationale that explains what evidence they used to construct their solution (SEPs 6, 7, and 8). In the process of discussing and writing this summary, group members are able to solidify their understandings. When groups are asked to complete a summary, individual scores on content tests are often higher than if the summaries are not used.

The following vignette section offers an example of how this assessment might be implemented in Ms. Sampson's lesson.

Ms. Sampson's Science Classroom: Group Summary of Solutions

Ms. Sampson asked each group to write and turn in a summary of the solution they had developed on the second day of the lesson. In these summaries, she noticed an issue that needed to be explained. Her students wrote that cold fronts make rain, and warm fronts lead to sunny weather.

Ms. Sampson: Alright, kids, I saw in your solutions that many of you wrote that only a cold front creates rain and storms. And nobody mentioned what happens if two air masses move toward each other, and they both have the same "strength." Does anyone know what a stationary front is?

The students looked confused, so she asked them to view an online computer simulation that let them see animations of four different kinds of fronts. She gave groups time to try out each kind of front and take some notes.

Ms. Sampson: So what did you find out about fronts?

Sarah: This shows how each kind of front can make it rain.

Carlos: Yeah, and that stationary fronts make big storms.

Jeremy: So do occluded fronts. Did you see that big cloud that formed in that part of the simulation? I bet that makes a huge storm!

Mai: Yeah! My dad says that when a cloud makes that flat place on top, it's a really big thunderstorm.

Ms. Sampson: OK, those are some good observations. How do you think this relates to your weather maps?

Jason: It means we have to know about how strong an air mass is before we can predict the weather.

Angie: And it gives us a way to know if there is going to be a severe storm. We need to look at temperatures, and pressures, and winds, and patterns on the map. It's kind of complicated, but it's kind of important, too.

Ms. Sampson: That's a great comment! Now, let's go back to the computers and look at the simulation again. Let's see if we can use the shapes of clouds to identify what's happening in our weather right now.

Summary

Facilitating PBL requires a slightly different set of skills from direct teaching, and it requires practice. Your role as the facilitator means you need to be prepared for several possible paths students may take. Your role also shifts from a provider of information to a guide who needs to skillfully ask questions that allow students to reveal their own thinking, resolve their own misconceptions, and base their own ideas on evidence rather than an "expert" source. This questioning also requires you to moderate disagreements and keep students on task, so facilitating PBL lessons will feel very different from other lesson formats.

You will also need to anticipate what kinds of information, models, and explanations you should be ready to offer your classes. If you teach multiple sections of the same class, each may have very different needs, so you will find yourself selecting different responses. Assessment is a key factor; you need to know what your students are thinking!

Box 3.2 presents some tips to remember as you facilitate your PBL lessons.

Box 3.2. Dos and Don'ts of PBL Facilitation

Do ...

- Use open-ended prompting questions.

- Count to 10 or 20 before making suggestions or asking questions.

- Allow learners to self-correct without intervening.

- Be patient and let learners make mistakes. Powerful learning occurs from mistake making. Remember that mistakes are okay.

- Help learners discover how to correct mistakes by clarifying wording, seeking evidence, or checking for discrepancies between ideas and evidence.

Don't ...

- Take the problem away from the learners by being too directive.

- Send messages that they are thinking the "wrong" way.

- Give learners information because you're afraid they won't find it.

- Intervene the moment you think learners are off track.

- Rush learners, especially in the beginning.

- Be afraid to say, "That sounds like a learning issue to me" instead of telling them the answer.

- Rephrase learners' ideas to make them more accurate.

Source: Adapted from Lambros 2002.

References

Bandura, A. 1986. *Social foundations of thought and action: A social cognitive theory*. Upper Saddle River, NJ: Prentice Hall.

Cornelius, K. E. 2013. Formative assessment made easy: Templates for collecting daily data in inclusive classrooms. *Teaching Exceptional Children* 45 (5): 14–21.

Dinsmore, D. L., P. A. Alexander, and S. M. Loughlin. 2008. Focusing the conceptual lens on metacognition, self-regulation, and self-regulated learning. *Educational Psychology Review* 20 (4): 391–409.

Lambros, A. 2002. *Problem-based learning in K-8 classrooms: A teacher's guide to implementation.* Thousand Oaks, CA: Corwin Press.

National Research Council (NRC). 1996. *National Science Education Standards.* Washington, DC: National Academies Press.

National Research Council (NRC). 2000. *Inquiry and the National Science Education Standards.* Washington, DC: National Academies Press.

NGSS Lead States. 2013. *Next Generation Science Standards: For states, by states.* Washington, DC: National Academies Press. *www.nextgenscience.org/next-generation-science-standards.*

Schwartz, D. L., C. C. Chase, and J. D. Bransford. 2012. Resisting overzealous transfer: Coordinating previously successful routines with needs for new learning. *Educational Psychologist* 47 (3): 204–214.

Zhang, M., M. Lundeberg, T. J. McConnell, M. J. Koehler, and J. Eberhardt. 2010. Using questioning to facilitate discussion of science teaching problems in teacher professional development. *Interdisciplinary Journal of Problem-Based Learning* 4 (1): 57–82.

USING PROBLEMS IN K–12 CLASSROOMS

In Chapters 1–3, we described the design of problem-based learning (PBL) lessons, shared tips for facilitating PBL activities, and gave you a taste of what PBL looks like in the classroom setting. But you may still have questions about when to use PBL, how to integrate PBL into the rest of your curriculum, and how to manage students and groups as they work through the problems you present. In this chapter, we will share some tips based on our own experiences and on the experiences of other K–12 teachers in the PBL Project for Teachers. These tips will help you plan to integrate PBL into your curriculum, identify potential resources, weave existing hands-on and inquiry activities into the PBL lesson, and assess student learning.

Establishing Continuity Across Lessons

Many classroom teachers like to establish a consistent set of routines or procedures for students to follow. This can help students learn habits of organization, acceptable ways to communicate, and skills needed to succeed in the classroom. If your classroom is built on specific procedures, PBL may feel very different from the "usual" routine. Students may take some time to learn the PBL process and how to think about problems in the PBL framework.

One of the benefits of PBL is that it helps students create the same kinds of productive habits of mind used by scientists in most fields. The three-category analytical framework used to discuss Page 1 and Page 2 creates a structure that students can apply to a wide range of different problems and content subjects.

But what if the PBL framework does not match up with your routines? Perhaps you like to start a unit with vocabulary and discuss concepts before doing an inquiry activity. Or maybe you like to start with a lab, discuss concepts, and then apply the idea to real-world problems. Many teachers have a set structure for phrasing hypotheses or an organization scheme for recording information in a lab notebook or science journal during labs.

For all of these examples, PBL can be incorporated as part of the routine. In fact, learners quickly adopt the analytical framework as a standard way to solve problems. PBL is also flexible enough to let you integrate other routines into the PBL framework. For instance, your students can express their hypothesis statements during the PBL analysis using an established format you use with other investigations. If journals are a part of your classroom routine, you can have students record their ideas in their journals as their groups discuss a problem. Or you may wish to have students post their "need to know" questions on a class wiki or cloud storage space to be answered as they find pertinent information.

We believe the key is to be consistent in some of the aspects of your teaching. As a professional, you can make choices about which procedures to maintain. You have the ability to modify parts of the PBL process to fit your teaching style and preferred routines.

It is possible you may opt to use PBL to make major changes in your teaching practice. Or you may be looking for new strategies to supplement your units. In the following sections, we describe some issues and strategies related to modifying PBL to fit within your curriculum.

Incorporating Other Activities

As you consider adopting problems in this book, you may be reluctant to give up the related learning activities you currently use. Maybe you have a great demonstration to model the phases of the Moon using lamps and balls held in different positions. If these are effective lessons, you probably don't want to toss them out to implement a new activity from this book.

Some critics of PBL suggest that the problems presented in this format are not suited to hands-on investigations. Some of the problems described in this book are based on finding information from text and online resources, so this may be the case for some problems. But we have also included activities with investigations. We encourage you to insert your best inquiry activities wherever and whenever they fit. PBL *can* be inquiry-based teaching! As you read through this chapter, you will see some examples and possible strategies for doing that.

Let's look at the E.T. the Extra-Terrestrial problem from Chapter 8 of this book. The story asks students to compare the changing phases of the Moon in a popular movie to the actual time frame of the phase changes to see if the movie shows the Moon accurately. The students look at still photos from the movie with a timeline from the movie's story. They compare this with the 28-day cycle of the Moon to see if the movie scenes are plausible.

Modification Ideas for E.T. the Extra-Terrestrial Problem

The story presents a scenario about an investigation to see if a popular movie's timeline accurately shows the changing phases of the Moon. The problem can be supplemented or modified using additional activities. Here are some modification ideas:

- Students can record a month-long log of Moon phases and times of moonrise and moonset to gather evidence of the Moon's movements.

- Students can use an online simulation or one of several apps for tablets that lets them view the phases of the Moon and its relative position for a month, a year, or a longer period of time.

Investigation in the "Research" Phase of PBL

After reading and analyzing Page 1 and Page 2, students should have a list of "need to know" issues. They will plan a strategy for searching for information, which is likely to include searching the internet. Students can find the information they need to solve this problem entirely online. But this may also be an ideal time to plan for an inquiry activity. Information from other sources is great, but primary information collected by doing an experiment or investigation is a powerful thing. Because one of our goals in teaching science is to help students use evidence to support their claims, doing experiments should be part of your science teaching. You can then help students relate the evidence they collect to the information found in online searches or from texts and construct solutions to the problem that reflect the conclusions drawn from these investigations.

A likely fit in the E.T. problem would be to engage students in a long-term observation study of the Moon's phases. The teacher may want to have students collect data through direct observation of the Moon so they can construct their own model of the 28-day cycle. The teacher can modify the lesson to introduce the story, and then help students make daily records of the Moon's phases and the times when it rises and sets. The data could include photographs or drawings and could span a month or more. Many of the problems in Chapters 5–8 can be modified in similar ways.

Investigation as a "Teachable Moment"

Sometimes the lab activities you use make a great response to a "need to know" item that arises during a class discussion. Using the Keweenaw Rocks problem in Chapter 6 as an example, students are likely to notice that Michigan has igneous rocks, evidence of activity involving magma or lava. A student may ask, "If there were volcanoes, why aren't there any volcanoes on the map? That doesn't make much sense to me." This might be the right time for some "just in time" instruction. PBL is certainly flexible enough to allow you to take advantage of these moments. For the Keweenaw Rocks example, you could have students read some text or online resources to learn about shield volcanoes. Another possible activity would be to build a model of Mauna Kea (a Hawaiian volcano) and compare it to the topography of the Upper Peninsula of Michigan to see if there are similar characteristics.

These activities might already be part of your curriculum. Rather than replace them with a PBL problem, a little planning will let you integrate your existing activities with the new PBL problems. In fact, giving students a chance to collect real data and use them with the PBL problem strengthens their understanding of how science is important to solving authentic problems.

Making Time for PBL and Timing PBL

Teachers often ask when PBL should be incorporated into an existing unit. Is it a way to introduce a topic, or should it be used to assess understanding at the end of a unit? Is it best used as an application task, or can students actually learn new concepts during a PBL lesson? The simple answer is … Yes to all of these! PBL is a strategy that is flexible. Figure 4.1 illustrates a number of ways in which PBL can fit in your unit plans.

Figure 4.1. Integrating PBL Into a Curriculum Unit

DIFFERENT SEQUENCES FOR INTEGRATING PBL INTO A SCIENCE UNIT				
PBL to engage learners	**PBL activity:** Engage students and introduce the relevance of a topic	Discuss solutions, reinforce concepts	Investigation— inquiry activity with concept	Student report or presentation
PBL to explore a concept	Pre-assessment of a concept	**PBL activity:** View the concept in context, build new ideas, confront misconceptions	Discussion for concept building	Summative assessment
PBL to apply a concept	Introduce a concept	Direct teaching— vocabulary and concepts	**PBL activity:** Apply the concept	Activities for practice, remediation	Test or project as summative assessment
PBL as a summative assessment	Introduce a concept	Lecture and reading from text	Practice— worksheets or sample questions	Lab activity	**PBL activity:** Use as application or summative assessment

Problems like those presented in Chapters 5–8 can be used in many different ways within a unit. Often, teachers think of PBL as a follow-up to concept-building activities or as a summary activity to assess students' ability to apply the concepts taught during a unit. And PBL certainly can be used in this way. As you read through the problems we have developed, you are likely to see clear contexts in which students can apply ideas you teach to other activities. When used at the end of a unit, students can demonstrate their ability to connect the new concepts to previous ideas or to integrate skills taught in other subjects. The groups' list of ideas generated during the analysis process ("What do we know?"

"What do we need to know?" and "Hypotheses") can reveal the learners' ability to think critically, recognize the concept within a context, and use vocabulary appropriately to talk about scientific principles, data, and their inferences.

But the end of a unit is not the only time when PBL is a useful learning tool. Teachers can use the problems as an "engagement" or motivation activity to introduce a topic. In this configuration, the story and challenge that serve as the focal point of a PBL problem paint an image that can spark students' interest in a topic. For instance, Earth's landforms and water strand (Chapter 7) includes stories about creating scientifically accurate scenery from across the United States and from below the ocean surface for a series of comic books. These lessons could be used to introduce a new topic while teaching Earth and space science standards. When you use PBL as an introduction to a concept, the final solution may be delayed until after you have taught some concept-building lessons, including the ones you probably already use in your unit. The purpose of PBL in this case is to provoke a "need to know" response from students so they will understand why they need to learn about landforms.

So PBL works as a summary assessment activity and as an introductory activity, but there is another important place for PBL in your curriculum. PBL problems can be used as concept-building activities. Many teachers do not think of the problems as a "main course" in the unit plan, often because they view PBL as an application activity that must follow introduction of vocabulary and concepts. But in our research and practice, we find that PBL can be an effective strategy for helping students learn new concepts.

The PBL problems build concepts in a different manner from direct teaching. Rather than *giving* information to students, a problem presents a reason for students to *find* or *construct* their own understanding. The analytical framework then helps scaffold the process of organizing their ideas and planning their search for new information. The final results are a product of students' pursuit of a solution to the problem that requires new understandings on their part.

This view of PBL as a concept builder also permits or even requires the teacher to incorporate experiments, investigations, and other experiences within the "research" phase of the PBL process. Your existing lab activities, homework, and practice exercises might be the ideal strategy for addressing something in the "need to know" list. The difference is that *students* have identified the need for those activities, rather than the teacher prescribing a required activity. This is a powerful tool for motivating students to engage in your existing lessons.

But does PBL work as a way to help all students learn new ideas? Some critics have questioned whether a class of students with mixed prior knowledge all benefit from PBL. Can students with no experience with a concept succeed in a PBL lesson? Do students who have a strong understanding of the concepts still learn in this format? These questions are important considerations, and the PBL Project for Teachers collected data to answer these questions. Our research showed that even with a diverse group of teacher learners, PBL helped 80.5% improve their content knowledge (McConnell, Parker, and Eberhardt 2013b). These results were seen with teachers who began the lesson with little or no prior content knowledge and others who entered with a high level of content knowledge. For teachers who are expected to address specific content standards, these findings suggest that they can successfully build conceptual understanding through PBL. This finding was bolstered by the fact that teachers in the project reported that PBL was effective in helping special needs students, high achievers, and the entire spectrum of learners in the classroom.

Assigning Individual and Group Work

PBL lessons naturally lend themselves to group work. Researching the "need to know" list, sharing proposed solutions, and carrying out investigations are all tasks that teachers usually plan to be carried out by groups of students. There are many grouping strategies that can be used within the PBL framework. Teachers should also consider how they will permit individuals to function in the PBL setting.

When your lessons reach a point when group work makes the most sense, you can use your normal procedures or policies for grouping. Our experience suggests that groups of four students seem to be most effective, although pairs can also function very well during the research phase. Groups of more than four may be unwieldy and may leave one or two students unoccupied with the tasks at hand. The general rule we recommend is to set group sizes at the number of students *needed* to complete the task. A lab may need groups of four or five. A simple online search may need two or three.

There are also numerous ways to assign students to groups: student choice, randomly drawn names, deliberate teacher-selected groups based on ability levels (heterogeneous or homogeneous), or groups based on who works well together. The most common strategies in the literature suggest that teachers should assign groups rather than allowing students to self-select (Johnson and Johnson 2008), but you need to select the strategies that fit your teaching style and classroom culture.

An Alternative Use of Grouping

In the typical PBL lesson, as described in the vignette about Ms. Sampson's classroom in Chapter 3, the initial discussions of Page 1 and Page 2 of the story take place in a whole-class setting. The students read the story and generate a shared list of ideas in each of the three categories ("What do we know?" "What do we need to know?" and "Hypotheses") of the PBL analytical framework. But there are opportunities to group students in this initial analysis of the problem as well. One of the teachers in the PBL Project for Teachers, whom we will call Kylie, was very successful in creating and employing this strategy.

After her eighth-grade science class read Page 1 of a PBL story, Kylie would ask students to work in groups of three or four to discuss the story and generate their own lists under the three categories of the framework. Groups were responsible for recording their ideas in a PBL journal and were given a time limit. When the allotted time ended, Kylie would call on one member of each group and ask them to list one item from one of their categories. As she recorded the ideas on the whiteboard, groups would continue adding ideas and discussing other groups' ideas until they had completed the first analysis. Then the class would read Page 2 and repeat the group analysis once again.

After the students researched "need to know" topics, each group was asked to present a proposed solution that was recorded on the whiteboard, and the groups then discussed the list of proposals. The final discussion included a vote by groups on the best solutions and a discussion of the science concepts related to the problem and solutions. This strategy was very effective in keeping every student engaged in the initial analysis.

Individual Accountability

As in many science activities, one of the challenges you face is that it can be hard to know what each individual thinks or does during group work. In any science activity, a group of four may have two students actively doing an experiment while the other two observe … or find other distractions that take their attention away from the learning activity. In a whole-class discussion, it might be hard to include each student's ideas. PBL lessons present the same challenges.

Just as in other lessons, your knowledge about the individuals in your class can help you adjust your grouping to better engage the students in your classroom who might withdraw or take a more passive role in group activities. Consider giving individual students responsibility for finding information about one of the learning needs the group has identified or using strategies to get each student to share at least one idea with the rest of the class. In the next section, we discuss strategies we have tested to help the teacher assess learning for each individual rather than relying only on a group presentation to evaluate every learner in the group.

Assessing Student Learning and Feedback to Students

Early in the lesson planning process, teachers need to consider how they will assess learning. This is an important task for any lesson format, including PBL. The PBL Project for Teachers provided us with opportunities to develop, experiment with, and revise assessment strategies and instruments for PBL lessons (McConnell, Parker, and Eberhardt 2013a). The result is a multifaceted assessment plan that was applied to each problem in Chapters 5–8 of this book. In those chapters, you will find open-ended general questions and application questions that can be used as pre- and post-assessments, transfer tasks to assess students' abilities to apply concepts, prompts for writing summaries of proposed solutions, and common beliefs inventories.

As part of our research and development, we tried several types of assessment questions. Some assessments were presented as multiple-choice "concept inventories" developed from national tests, but these did not give us much information about learners' ideas relating to *how* or *why* they chose their answers. We also tested concept maps as an assessment strategy, but these were difficult to evaluate because the learner was unable to describe the nature of the connections between ideas in detail. Interviews with individual learners can provide more insight into student thinking, but the time required to assess *every* student makes interviews impractical for most classroom settings.

The assessment questions that were most useful were open-response questions to which learners wrote their responses. Some teachers may find the responses to these open-ended questions very time-consuming to read, but no other type of question was effective in showing what ideas learners have and how they connect those ideas to each other. These connections were very revealing—they often exposed misconceptions and gaps in students' understandings.

In the rest of this chapter, we describe the various types of assessments and questions that may be used in a PBL lesson.

Pre- and Post-Assessments: General and Application Questions

As you teach a PBL lesson, it is important to assess what students know before you begin. For our research, we used a pre-problem assessment and an identical post-problem assessment to describe the new ideas learned and changes in teachers' ideas about science. You can do the same with your students. But developing questions that assess both the breadth and depth of your learners' ideas is complex.

In the PBL problems presented in Chapters 5–8, we have included the assessment questions we tested, revised, and retested. You will find two kinds of questions: *general questions* and *application questions*. Let's look at examples from Chapter 7, "Weather":

- *General question:* Explain why water does not disappear from the water cycle when an animal drinks water.

- *Application question* (from Problem 2, Water So Old): Explain how a water molecule could get from the bottom of the ocean to the top of Mount Everest. Your explanation should include what drives the water molecule's movements and changes.

So why two types of questions? Aren't they asking the same thing? At first glance, they both appear to ask about the same concept, but the wording can lead to very different types of answers that reveal very different levels of understanding!

For the example general question, our learners wrote brief explanations that included a general comment that water is given off by animals during respiration. Sometimes the responses were a string of words with arrows to show connections, as shown in the following examples:

> *Response:* Water is taken in by animals when they eat and drink. Then they exhale water given off during respiration, and return the water to the water cycle.

> *Response:* Precipitation $\longrightarrow$ Runoff $\longrightarrow$ Lakes $\longrightarrow$ Animals $\longrightarrow$ Respiration $\longrightarrow$ Water vapor

Both of these responses can be considered "correct," but what does it mean when students write these? The first response uses terms you might see in a textbook definition, but it appears the writer holds to the misconception that the water from respiration contains the same molecules the animals drank. The second response includes more terms but does not explain mechanisms at all. Does the learner understand the mechanisms but simply left them out of the response? The "descriptive" knowledge displayed here is important, but we want students to know more than just a definition. We'd like them to understand why and how water moves between reservoirs so that they can predict and explain real events they witness in their own lives.

In the example application question for the Water So Old problem, we describe a specific example that demonstrates the movement of a water molecule through the water cycle. The question then prompts the student to explain *the mechanism* that makes the water cycle work, as shown in this example:

> *Response:* The water molecule would have to be carried by a current to the surface of the ocean. The currents are driven by uneven heating of the ocean. Once on the surface, the water molecule could evaporate if it absorbed enough energy from sunlight. It would become part of a warm, moist air mass. When some kind of weather carries that air to the mountains, it snows because it is over land. The Himalayas would also make the air cooler. As the water molecule cooled, it would condense and precipitate as a snowflake that could fall on Mount Everest.

This response helps us see that the writer has an idea that ocean currents are caused by uneven heating, and that water molecules evaporate when they have absorbed enough energy. While this answer has some accurate information, it also reveals some incomplete or vague understandings. An accurate and complete description would describe why the air mass cools and would include the possibility that the water molecule could absorb energy from the warm water or air around it.

This application question requires the learner to *explain* rather than just describe or define. The written response to an application question is often longer than for a general question, and, as in this case, can show which *parts* of a concept the learner has mastered and which are not yet fully developed or accurate. In this case, the student can describe the pattern, but there is no evidence that she truly understands the mechanisms that cause lake-effect snow.

So why not just ask the application question? For many topics, learners tend to focus on a couple of key elements of an application question. The general question gives room for the student to add ideas that may be peripherally connected or to leave out concepts that are important. We found that if we first ask the general question, followed by one or two application questions, we gain insight into the overall awareness students have of a concept *and* how deeply they understand the idea. Together, these questions give us a great view of the writer's understanding of the forces that create weather patterns.

TEACHING TIP: Implementing the Two Types of Questions

Both the general questions and the application questions are a good way to assess what each *individual* student knows! We suggest you NOT use these as a group writing assignment.

In your classroom, you can use these questions as part of your own assessment plan. The questions are very well suited to be a pretest, either for a unit or for each specific PBL problem. These questions can also be used as post-assessment questions, either in the form of a quiz or open-write about the specific concept or as part of a chapter or unit test.

Along with each question, we have included a model response that you may use as a guide for developing a scoring key or a rubric for evaluating the responses.

The second sentence in the example application question ("Your explanation should include what drives the water molecule's movements and changes.") is also very important in getting a detailed explanation. In our first trials, we left that sentence out because we felt that this was leading the student to the answer. But without it, learners seemed to write incomplete answers. They might write about one important term and leave others out. In our later trials with the more specific prompt, we found that responses were more thorough but still included inaccurate or incomplete ideas and gaps in understanding. We also saw responses that overtly stated that the student did not know part of the response (e.g., "I'm not sure—there would be movement of the water vapor in the air. The clouds are 'fuller' at Buffalo, NY.").

Student Drawings

When you ask students to respond to the pre- and post-assessment questions, allow them to include drawings to illustrate their ideas. In fact, encourage it! Their drawings often reveal misconceptions and accurate understandings better than text. Some learners are verbal, but others are visual, and their conceptual understandings may be easier to communicate in a drawing.

As an example, let's look at a drawing (see Figure 4.2) that was a response to the following pretest question from the astronomy strand (Chapter 8) in the PBL Project for Teachers:

> Why do we see the Moon go through phases, and why do they change? Explain clearly in words and support your explanation with a well-labeled diagram.

The pretest question refers to a concept addressed in the E.T. the Extra-Terrestrial problem that helps students understand the cycle of Moon phases.

Figure 4.2. Sample Student Drawing for Pretest Response

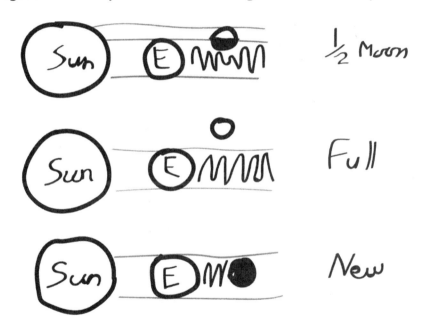

Michael: Sun/Earth/Moon alignment creates phases in the Moon. Earth's shadow is cast at different angles onto the Moon's surface.

The text and diagram in Figure 4.2 both suggest that the learner has a scientifically inaccurate view of what causes the phases of the Moon. He appears to understand that the light coming from the Moon's surface is reflected light from the Sun, but he believes that the "dark" portion of the Moon is created by Earth's shadow. The diagram and written explanation do not account for the angle of incidence. Instead, they suggest that sunlight illuminates the entire Moon. This is a common belief for the concept of Moon phases, and the teacher can use this information to identify the nature of the misconception and plan an activity that will help students see how the phases are produced, perhaps with an inquiry activity using a lamp, a ball, and different positions in the room to model the Earth-Moon-Sun system (SEP 2: Developing and Using Models; SEP 7: Engaging in Argument From Evidence; CC 1: Patterns; CC 2: Cause and Effect: Mechanism and Explanation; CC 4: Systems and System Models).

Transfer Tasks

Have you ever taught a concept, thought your students "got it," and then asked a question on a test or exam that represented the very same concept in a different context? Did your students fail to connect the two situations and recognize them as the same concept? All too often, the answer to that question is "Yes!"

The problem of getting students to transfer their learning to new situations is nothing new (Ellis 1965). One of the things we want to know about PBL is how well students can transfer what they learn as they resolve a problem to new contexts. In each of the content chapters in this book, we have included *transfer tasks*—open-ended questions that ask students to apply the concept taught in a problem or group of problems to a new situation. Let's look at an example of a transfer task from the Northern Lights problem in Chapter 7:

> **TEACHING TIP: Using the Transfer Tasks**
>
> Transfer tasks can be used in your lesson plan as a way to check understanding at the end of the PBL problem, either as a short stand-alone essay or as part of a chapter or unit test. This makes an excellent individual task to help you assess each student's learning.
>
> Along with each transfer task, we have included a model response that you may use as a guide for developing a scoring key or a rubric for evaluating the responses.

The table below shows the times of sunrise and sunset on the same day for three cities in the United States. Use the table to answer the following questions:

a. It is the time of year when temperatures are getting warmer. Approximately what time of year is it? Explain your answer.

b. Which city is farthest north? Which city is farthest south? Explain your answers.

In the Northern Lights problem, students explore the patterns in length of daylight and the path of the Sun in the sky associated with the seasons (CC 1: Patterns). In the transfer task, students need to think about the patterns that they learned in the same way as in the story presented in the problem, but now they are applying the concepts to a new context. The model response included in the "Assessment" section of this problem gives some ideas about the reasoning students could be expected to use.

Solution Summaries

Another important source of information about what your students have learned will be the solutions they develop as a final product of a PBL lesson. Individuals or groups build these solutions as an answer to the "challenge" presented on Page 1 of the problem. In some cases, the solutions may be brief, but in other problems, the solutions may be an extensive paper, a model, or even a demonstration. This type of assessment is flexible enough to be used as an individual assessment or a small-group assignment to help plan for discussions.

In Kylie's classroom, each group was required to share their solution and an explanation of why the group members thought their solution was correct. The class then voted for the solutions that made the most sense. This was not a way to select "the right answer," but it served to support Kylie's effort to foster discussion and evaluation of which solutions best matched the information collected. In many cases, there can be more than one correct solution, so a consensus agreement is not needed.

In our own use of PBL, this assessment usually entailed having individual learners write a page or two about their resolution of the problem. The benefit of an individual written solution is that it gives the teacher a way to assess the understanding of each individual in the classroom. The individual assessments are important because much of the sharing of ideas throughout the analysis of the problem will have been done in groups. It is possible, even likely, that the teacher will hear many correct statements, but they come in bits and pieces from individual students. Some students may not have shared their ideas and questions, and others may have mastered one element of a complex problem or concept, but this "group speak" view of the understanding of the class as a whole does not necessarily translate to a complete understanding for each individual.

By asking each student to summarize the final solution to the problem, the teacher can find out which students have been able to connect concepts and apply them accurately and completely and which students still have gaps in their understanding, have misconceptions, or are describing concepts in vague or "fuzzy" language. In some cases, the solutions will point to a need for the teacher to lead the class in additional activities, explain ideas to the whole class, or ask some probing questions in a discussion.

When the assessments are used by the teacher to guide instructional decisions as described here, PBL becomes a rich context in which to frame assessment questions. The PBL lesson is not the only activity for teaching concepts in the unit, but it is a key component that we suggest gives the teacher a far better understanding of the depth and limits of what learners actually know and learn.

Let's look at an example from the Keweenaw Rocks problem in Chapter 6. In this problem, students learn about the geologic history of the Keweenaw Peninsula, a region that has an unusual collection of igneous rocks mixed with sedimentary and metamorphic rocks. The following is a solution summary written by a learner in the PBL Project for Teachers:

Over 1 billion years ago there was volcanic activity in the middle of Lake Superior. This volcanic activity occurred because of a rift that formed along a plate boundary. The upward movement of magma caused the crust to split and magma came to the surface. This is the basalt we have on the Keweenaw Peninsula. This "erupting" or rifting occurred over a long period of time, about 25 million years. There was a lot of time between eruptions. In the time between eruptions, sandstone and conglomerates were deposited on top of the lava layers by streams and temporary lakes. The lava spread out, the sedimentary rocks formed on top, at some point the boundary became convergent and caused reverse faults which pushed the basalt (lava that has cooled), and sandstone up, forming the Keweenaw Peninsula and Isle Royale.

The solution summaries provide important evidence that teachers need as they teach a PBL lesson. They reveal learning outcomes, identify gaps in students' understandings, and help the teacher to recognize whether students have assimilated ideas discussed in the group discussions or found in the research phase of the PBL process. For instance, the part of the response that describes how the basalt formed at a rift zone reveals some important learning. The writer also included an accurate response about the formation of sedimentary rocks. But it also shows that the learner may not have learned that the boundary where sedimentary rocks came in contact with new magma or were compressed in the uplift of the basalt and sandstone results in metamorphic rocks found in the Keweenaw area.

The information gaps or misconceptions in students' understanding gained from these written summaries of the solution are important tools in helping to assess learning and plan the next steps in instruction and planning. But there are also tools available during the lesson that can provide even more data to help direct instructional decisions. These tools include the formative assessments discussed in the following subsections.

Formative Assessments

The assessments described earlier are based on questions deliberately posed to learners as either a diagnostic or summative assessment. But like any teaching strategy, PBL presents many opportunities to conduct formative assessments of student thinking and learning. In fact, much of the structure of the PBL analytical discussion gives the teacher a clear picture of how students' ideas are changing. Formative assessments also allow the teacher to adjust, adapt, and revise plans on the fly. The following subsections describe some parts of the PBL process and some simple strategies the teacher can add that act as formative assessments.

PERFORMING AN INITIAL ANALYSIS

In earlier chapters, we suggested that the facilitator record ideas generated during the three-part analytical process ("What do we know?" "What do we need to know?" and "Hypotheses") on paper, a whiteboard, an interactive whiteboard, or some other form of display. The entire discussion of the story (Page 1 and Page 2) involves students sharing ideas, asking questions, and creating hypotheses. Each of these statements is a valuable bit of information for the teacher.

As you lead your class through a PBL lesson, it is important that your list of ideas be kept in some format you can store. Students nearly always change their ideas or revise hypotheses, and as they do this, you can see evidence of learning. By keeping each idea and either crossing out or modifying them, your permanent record helps you trace this development. Even your students will see and talk about how their ideas have changed.

We have also found that the lists help you identify opportunities to pause the PBL process long enough to answer a question or correct a misconception. You may also find that students' ideas and questions lead you to a need to do certain hands-on investigations or demonstrations as part of the instructional process. Without a clear record of student-generated ideas, you may have a hard time catching those "teachable moments" or bringing your class back on track when the pause is finished.

REPORTING RESEARCH FINDINGS

After the initial analysis, the next opportunity to assess the development of students' ideas happens as the groups research ideas listed in the "need to know" list and share their findings with the class. Some teachers use graphic organizers to help students record and report information they find from text and online sources. These documents can be a source of information in your formative assessment plan.

But you can also implement your own plan to have students put their information on the board to share with the class or give a short presentation of their findings. As they share what they have found, you can collect and record their ideas as more evidence of learning or to identify emergent learning needs.

Strategies for keeping this information may include science notebooks or journals, a presentation paper or poster, electronic whiteboard presentations saved on the teacher's computer, or videotaped records of groups' presentations. The only limit on how the sharing of findings can be done is the imagination of teachers and students. We recommend that you adapt the procedures you may already use when you ask students to share in front of the class.

CHAPTER 4

USING EXIT TICKETS

PBL lessons are often a process that takes more than one class period, especially if you teach in 45- to 50-minute periods. A typical plan might include the initial analysis (Page 1 and Page 2) on the first day, followed by a class period to research information, and a third session for presentation and final discussion.

The end of each day presents another opportunity for formative assessment. Many of the facilitators in the PBL Project for Teachers used exit tickets (see description and vignette in Chapter 3, p. 42) as a way to get a brief look at what students are learning or what topics they are struggling to understand. Examples of exit tickets included statements that asked students to write reflections that give the teacher feedback about the lesson, such as "What solution do you think is most useful?" "What is one topic you are still confused about?" and "One thing you've learned is …" The teacher/facilitator can review these short anonymous assessments to make choices about what topics to address with students as the next class period begins or how to help students understand difficult concepts.

Several books and websites offer ideas for these types of formative assessments. Examples of publications are *Seamless Assessment in Science* (Abell and Volkmann 2006), *Formative Assessment Strategies for Every Classroom* (Brookhart 2010), and "Formative Assessment Made Easy" (Cornelius 2013).

WRITING SUMMARIES OF THE "BIG IDEAS"

A variation of the exit ticket strategy that we tested in the PBL Project for Teachers was to ask learners to write a summary of the "Big Ideas" they learned from the PBL lesson. The Big Ideas are the foundational concepts that Roth and Garnier (2006) suggest teachers be mindful of when they plan and implement lessons. The lists of Big Ideas were generated by the participants in each content strand after they had reached the final resolution of a problem. Listing the Big Ideas they felt they had learned in the problem gave learners a chance to think metacognitively about the new ideas they were developing and focused their attention on the objective or learning goals of the lesson.

Our use of this assessment with teachers showed that sometimes we think learners are gaining knowledge of our target learning goals, but, in fact, they are thinking about other aspects of a problem. Because we want to know what the real outcomes are, these written summaries help us focus on the real changes in learners' thinking. Like the solution summaries described earlier, these Big Idea summaries often point to opportunities for other instructional activities. Again, the assessment guides our next steps as we teach the concept.

ADMINISTERING COMMON BELIEFS INVENTORIES

A final form of assessment that we tested in many of the problems represented an attempt to see if PBL lessons could help correct inaccurate or unscientific understandings. As you have probably noticed, there are many science concepts for which people have preconceived notions, some of which are not accurate. We call these common beliefs. The common beliefs inventories that we developed were tested to see if they were able to help identify misconceptions and to see if students' ideas about these concepts changed as they completed a PBL lesson. These assessments were not created for every problem, so we have included only a few examples in this book.

The common belief inventories were designed to be a brief measure to see what students think about common ideas and misconceptions. The inventories included about 10 true/false statements, and students were instructed to explain *why* each statement is true or false. This explanation is important because it probes more deeply into learners' ideas and avoids the trap of allowing an unexplained "guess."

The inventories proved helpful in some problems as a pre- or post-assessment. In the PBL Project for Teachers, we administered the common beliefs inventories at the beginning of the workshop and had participants revisit the questions after each problem was solved to see where new information might change their responses. The results showed instructors which common beliefs had changed and which remained consistent even after the lesson. In some strands, the inventories were given multiple times, and learners were asked to compare their own answers on each trial with previous tests. Letting students look back at their own learning is a powerful form of metacognition that helps them become more reflective and self-regulating.

As an example of the information the teacher can gain from a common beliefs item, let's look at an assessment from Chapter 7. One of the true/false statements in the weather Common Beliefs Inventory reads, "The Sun's light is very steady and uniform, but Earth is not heated uniformly by it." This simple statement is accurate; the energy coming from the Sun is constant, and the amount of energy that reaches the surface of Earth is steady. But Earth is not heated evenly, because regions closer to the poles receive light at a more indirect angle and because some areas have different materials that absorb or reflect sunlight differently. Let's look at some responses to this statement listed by learners in the PBL Project for Teachers:

- **Lisa:** False: Some parts of Earth don't get the same amount of sunlight because of the seasons.

- **Sharon:** False. The Sun is more concentrated in some areas, so those places heat up more.

- **Robert:** True. The same amount of sun hits every part of Earth, but places with more clouds or closer to the poles do not heat up as much.

Not only do the true/false answers vary, but the reasons behind their choices differ dramatically. Lisa believes that seasons cause uneven heating, conflating cause and effect. Robert's answer has the correct true/false designation, but it is incomplete because he doesn't mention uneven heating due to differences in the directness of the Sun. Sharon's response is difficult to interpret. The teacher may need to ask follow-up questions to determine what she means by "concentrated." Students' statements that are difficult to interpret often reflect confusion on the part of the student. The examples, models, and explanations a teacher chooses for a group of learners need to take into account the specific ideas underlying the students' ideas, and the common beliefs inventories give yet another tool for assessing those ideas.

◉◉◉

As you explore the various problems in Chapters 5–8, look for the different assessment questions and the model responses. You are welcome to use the assessments as we present them, but you may also wish to modify and adapt the questions.

References

Abell, S. K., and M. J. Volkmann. 2006. *Seamless assessment in science: A guide for elementary and middle school teachers*. Portsmouth, NH: Heinemann Educational Books.

Brookhart, S. M. 2010. *Formative assessment strategies for every classroom: An ASCD action tool.* Alexandria, VA: Association for Supervision and Curriculum Development.

Cornelius, K. E. 2013. Formative assessment made easy: Templates for collecting daily data in inclusive classrooms. *Teaching Exceptional Children* 45 (5): 14–21.

Ellis, H. C. 1965. *The transfer of learning.* London: Macmillan.

Johnson, D. W., and R. T. Johnson. 2008. *Cooperative learning.* Hoboken, NJ: Blackwell.

McConnell, T. J., J. M. Parker, and J. Eberhardt. 2013a. Assessing teachers' science content knowledge: A strategy for assessing depth of understanding. *Journal of Science Teacher Education* 24 (4): 717–743.

McConnell, T. J., J. M. Parker, and J. Eberhardt. 2013b. Problem-based learning as an effective strategy for science teacher professional development. *The Clearing House* 86 (6): 216–223.

Roth, K., and H. Garnier. 2006. What science teaching looks like: An international perspective. *Educational Leadership* 64 (4): 16–23.

EARTH'S LANDFORMS
AND WATER

The problems in this chapter focus on the features of the surface of Earth. Students use topographic and relief maps to identify features on land as well as on the ocean floor. They also identify where water, both fresh and salty, is found. The problems involve students with crosscutting concepts (CCs) described in the *Next Generation Science Standards* (*NGSS*; NGSS Lead States 2013); these concepts include Patterns (CC 1), as they compare and contrast different landforms in different areas, and Energy and Matter: Flows, Cycles, and Conservation (CC 5), as they explore the water cycle. The problems also help students develop the scientific practices of Developing and Using Models (*NGSS* science and engineering practice [SEP] 2) and Obtaining, Evaluating, and Communicating Information (SEP 8).

The problems in this chapter include extensions that lead middle school students into the relationships between plate tectonics, water, and Earth's features. These extensions relate to the CC of Cause and Effect: Mechanism and Explanation (CC 2).

The disciplinary core ideas (DCIs) addressed in this chapter include The History of Planet Earth (ESS1.C), Earth Materials and Systems (ESS2.A), Plate Tectonics and Large-Scale System Interactions (ESS2.B), and The Roles of Water in Earth's Surface Processes (ESS2.C).

Big Ideas
Landforms (Grades 3–5)

- Topographic and relief maps of Earth's surface show patterns (CC 1: Patterns). On land, there are *mountains*—high, uneven places where the surface has been broken or folded and piled up. Mountains usually occur in groups or chains. The classic cone shape of volcanoes often shows on topographic maps as nested, closely packed, irregular circular contour lines. Rivers (and glaciers) form valleys as they flow downhill. *Canyons* are valleys with steep sides. *Plains* are large flat areas, and *plateaus* are flat areas that are higher than some of the surrounding areas. Small series of ridges that are lower and less uneven than mountains are called *moraines* when they are made of gravel and boulders left behind by retreating glaciers.

- Many of the same features, such as mountains, are present on the ocean floor as well as on land. Other features are easier to identify on the ocean floor than on land or are unique to the ocean. *Mid-oceanic ridges* appear as long, raised formations with many perpendicular fault lines crossing them. *Trenches* are very deep valleys often found off of the coast of continents. *Shelves* are shallow areas just off the coast of some continents or land masses.

Plate Tectonics (Grades 6-8)

- When topographic maps are compared with maps of the tectonic plates and maps that show earthquakes and volcanic activity, correlations are visible (CC 1: Patterns). Mid-oceanic ridges, trenches, and subduction zones are associated with tectonic and volcanic activity at the edges of plates. Mid-oceanic ridges occur where oceanic plates are diverging and magma wells up to fill the void. On land, these appear as rift valleys.

- Oceanic plates are made of dense rocks like basalts, while continental plates are mostly made of less dense rocks like granites and sedimentary rocks. When a dense oceanic plate collides with a less dense continental plate, the denser oceanic plate is subducted—that is, it is pushed under the continental plate. As the dense oceanic plate is pulled down by gravity, a trench forms at the margin of the continent. The continental plate deforms, crumples, and piles up, forming a mountain range parallel to the trench. The mixture of rock and water in the oceanic plate forms magma as it is subjected to heat and pressure. This magma sometimes erupts in a volcano. Thus, a colliding oceanic/continental boundary is often marked by a line of volcanoes and a trench.

- Two colliding continental crusts result in mountain formation as the crust at the boundary is broken or deformed and piles up. Mountains are subject to weathering and erosion from water and wind so that the uplifting process of mountain building is opposed by the destructive processes of erosion. Once uplifting stops, mountains are slowly eroded away. Thus, the degree of erosion is an indicator of a mountain range's age.

Water (Grades 3-5)

- Water exists on Earth in all three states of matter and is distributed in many places. Liquid water can be found on the surface of Earth, underground, and in the atmosphere as clouds and rain. Solid water (ice) is found in glaciers and polar ice caps or in the atmosphere as snow or hail. Water vapor is found in the atmosphere.

- Liquid water can be either fresh (lakes, rivers, groundwater) or salty (oceans). Almost all the water (~97%) on Earth is found in the oceans and is salty. Two percent of Earth's water is in the polar ice caps and glaciers. Less than 1% of the water on Earth is freshwater in the liquid state, and it is not evenly distributed over Earth's surface.

Water (Grades 6-8)

- Liquid water under the force of gravity weathers and erodes rock. It dissolves the rock or breaks it into tiny bits, which the water carries away. Thus, water (and wind) tends to flatten out the surface of Earth, while tectonic forces build it up. Water erodes valleys. It may also form mesas and buttes if it leaves harder rock formations behind.

- Liquid and solid water move or fall downhill because of the force of gravity. When water is heated directly or indirectly by the Sun, it may evaporate. This is how water moves up into the atmosphere.

Conceptual Barriers

Students are used to looking at the surface of geologic formations rather than seeing them as extensions of what is below the surface or manifestations of processes that shape Earth. In addition, geologic processes are slow. They take place over long periods of time and across vast areas, so it is difficult for students to visualize the events that produce the landforms we see on Earth's surface.

Common Problems in Understanding

- Many of the visualizations of the three-dimensional (3-D) geologic processes are two-dimensional and static.

- Students often draw simplified water cycles with only evaporation, condensation, and precipitation, ignoring the movement of liquid water and the presence of groundwater.

Common Misconceptions

- Earth's crust is one solid mass.

- Magma comes from Earth's core.

- People have drilled to Earth's core.

- Places that have cold seasons now have always been cold. Places that are above sea level have always been so. Places that are deserts now have always been deserts.

- Continents don't move and change.

- Rocks don't change.

- Earth is only a few thousand years old.

- All water follows the same path through the water cycle.

- As depicted in standard diagrams of the water cycle, it only rains over land and water only evaporates from oceans.

Interdisciplinary Connections

There are a number of interdisciplinary connections that can be made with the problems in this chapter. For example, connections can be made to literature, language arts, geography, history, mathematics, and art and music (see Box 5.1).

Box 5.1. Sample Interdisciplinary Connections for Earth's Landforms and Water Problems

- **Literature:** Read one of several books set in rivers and canyons and relate the story to landforms and the forces that create them. Examples include *Downriver* by Will Hobbs (2012), *The Great Lakes* by Sara St. Antoine (2005), and *The Legend of the Petoskey Stone* by Kathy-jo Wargin (2004).

- **Language arts:** Write a descriptive story about the forces that produced a prominent local landform or the way people might have lived before those landforms were created, or write a story about exploration of a deep open trench in a submarine.

- **Geography:** Use maps to identify places where various landforms can be found, or examine patterns of human uses of land around various landforms.

- **History:** Read about and discuss important historical events that were influenced by geologic landforms like mountains, volcanoes, or rivers.

- **Mathematics:** Use rates of formation of new land at a volcano to estimate the time it would take a seamount to reach the surface, or use rates of erosion to estimate the time it takes to form a canyon or mesa.

- **Art and music:** Examine art at a local museum that features various landforms; paint or sculpt a chosen landform; write a poem or a song about a canyon, a dune, or a mountain; or illustrate the response to the superhero adventure transfer task in the Water, Water Everywhere problem.

References

Hobbs, W. 2012. *Downriver*. New York: Simon & Schuster.

NGSS Lead States. 2013. *Next Generation Science Standards: For states, by states*. Washington, DC: National Academies Press. *www.nextgenscience.org/next-generation-science-standards*.

St. Antoine, S. 2005. *The Great Lakes: A literary field guide*. Minneapolis: Milkweed Editions.

Wargin, K. 2004. *The legend of the Petoskey Stone*. Ann Arbor, MI: Sleeping Bear Press.

Problem 1: An Eagle's View

Alignment With the *NGSS*

PERFORMANCE EXPECTATIONS	• *4-ESS2-2:* Analyze and interpret data from maps to describe patterns of Earth's features. • *MS-ESS2-2:* Construct an explanation based on evidence for how geoscience processes have changed Earth's surface at varying time and spatial scales.
SCIENCE AND ENGINEERING PRACTICES	• Developing and Using Models • Analyzing and Interpreting Data • Constructing Explanations and Designing Solutions • Obtaining, Evaluating, and Communicating Information

DISCIPLINARY CORE IDEAS	• *ESS2.A: Earth Materials and Systems*
	
	  ○ *Grades 3–5:* Rainfall helps to shape the land and affects the types of living things found in a region. Water, ice, wind, living organisms, and gravity break rocks, soils, and sediments into smaller particles and move them around.
	
	  ○ *Grades 6–8:* All Earth processes are the result of energy flowing and matter cycling within and among the planet's systems. This energy is derived from the Sun and Earth's hot interior. The energy that flows and matter that cycles produce chemical and physical changes in Earth's materials and living organisms.
	
	• *ESS2.B: Plate Tectonics and Large-Scale System Interactions*
	
	  ○ *Grades 3–5:* The locations of mountain ranges, deep ocean trenches, ocean floor structures, earthquakes, and volcanoes occur in patterns. Most earthquakes and volcanoes occur in bands that are often along the boundaries between continents and oceans. Major mountain chains form inside continents or near their edges. Maps can help locate the different land and water features of Earth.
	
	  ○ *Grades 6–8:* Maps of ancient land and water patterns, based on investigations of rocks and fossils, make clear how Earth's plates have moved great distances, collided, and spread apart.

CROSSCUTTING CONCEPTS	• Patterns • Cause and Effect: Mechanism and Explanation • Systems and System Models

Keywords and Concepts

Landforms, geologic forces, plate tectonics

Problem Overview

Students help a comic book publisher create the scenery and story lines that include real geologic formations in different locations on Earth.

Images available in full color on the Extras page (*www.nsta.org/pbl-earth-space*) are marked with the following icon: ✪.

PROBLEM 1

Page 1: The Story

An Eagle's View

The Cosmic Comic Book Company has just hired your class. The company is planning several new superheroes. One of these is Eli Eagle. Eagles are big birds (see Figure 5.1), but Eli is unusually big. His wingspan is 20 feet. He can fly up to 500 miles per hour (as fast as a jet airplane). His eyesight is very good. He can see animals on the ground even when he is flying 1,000 feet in the air at top speed.

✪ Figure 5.1. Bald Eagle

The comic book company wants to have exciting scenes of Eli flying over all kinds of different landforms. The company will use real maps (see Figure 5.2 for an example of a topographic map) and a computer to make the pictures of the landscapes. For example, a writer could say, "Scene with Eli flying over gently sloped mountains." The animators (people who create the pictures) would choose a map of an area with gently sloped mountains. The computer would then make a background picture using the map. But the writers need to know what types of landforms are found in the United States. Your class has been hired to answer this. Your boss suggests you start by looking at the Grand Canyon, a mountain range, and Wisconsin.

Your Challenge:

- *Make a list of types of landforms found in the United States. Describe each type of landform and explain where that type of feature can be found.*

- **Grades 6–8 extension:** *Describe and explain which landforms are associated with different tectonic boundaries.*

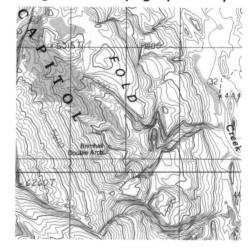

Figure 5.2. Topographic Map

PROBLEM 1

Page 2: More Information

An Eagle's View

Your boss lives in Oshkosh, Wisconsin, but he used to live in Phoenix, Arizona. He got the idea for the character Eli when he visited the Grand Canyon in the northern part of Arizona (see Figure 5.3). He was surprised that all of the "mountains" around the can-

yon were the same height. He was used to mountains being all different heights. He was also impressed by the red color of many of the rocks in the canyon and the fact that it is a mile deep in places.

✪ Figure 5.3. Grand Canyon

Your boss often flies from his home in Wisconsin back to Phoenix. From the air, he sees some very tall mountains. These are the Rocky Mountains. As he gets closer to Phoenix, he sees a flatter area. Then there are a few more mountains and then Phoenix. He says that Phoenix is in a flat area with small mountains around it. On the other hand, Wisconsin is much flatter. But in seemingly random places, there are hills called moraines.

Your boss is wondering what other landforms are in the United States. In order to generate a lot of stories about Eli, he wants to know about all of the different landform types. He would like you to organize the topographic maps that the computer will use by landform.

Your Challenge:

- *Make a list of types of landforms found in the United States. Describe each type of landform and explain where that type of feature can be found.*

- **Grades 6–8 extension:** *Describe and explain which landforms are associated with different tectonic boundaries.*

PROBLEM 1

Page 3: Resources

An Eagle's View

Topographic and Relief Maps

- Free topographic maps of most places in the United States can be downloaded in PDF format from the U.S. Geological Survey at *http://store.usgs.gov/b2c_usgs/ usgs/maplocator/%28xcm=r3standardpitrex_prd&layout=6_1_61_48_1&uiarea=2&cty pe=areaDetails&carea=%24ROOT%29/.do*.

- Free shaded relief maps of the United States are available at *http://birrell.org/ andrew/reliefMaps*. You can also do an internet search on the term *relief maps of the United States*.

- Google Earth (*www.google.com/earth*) allows you to fly over the surface of Earth.

Maps of Tectonic Plates (for Grades 6–8)

- See, for example, *https://planetjan.wordpress.com/2010/04/04/its-the-end-of-the-world-as-we-know-it/map_plate_tectonics_world*, or search online images for *map of tectonic plates*.

Books

Barker, C. F. 2005. *Under Michigan: The story of Michigan's rocks and fossils*. Detroit: Wayne State University Press.

Cole, J. 1987. *The magic school bus: Inside the earth*. New York: Scholastic.

Dorr, J. A., and D. Eschman. 1970. *Geology of Michigan*. Ann Arbor: University of Michigan Press.

National Wildlife Federation. 1997. *Geology: The active earth*. New York: McGraw-Hill.

Wargin, K. 2004. *The legend of the Petoskey Stone*. Ann Arbor, MI: Sleeping Bear Press.

PROBLEM 1

Teacher Guide

An Eagle's View

Problem Context

In this problem, students use topographic maps to find patterns in U.S. landforms. Although this problem asks students to look at landforms in Arizona and Wisconsin as well as a mountain range, any two or more locations with different topography could be used. The goal is for students to identify a range of landforms by comparing two or more sites.

Problem Solution

Model response for grades 3–5: Mountains are high, uneven places—higher and steeper than hills. The Sierra Nevada Mountains, the Cascade Range, and the Rocky Mountains are the major mountain ranges in the western United States. The Appalachian Mountains are the major mountain range in the East. Volcanoes are most often seen as single, cone-shaped mountains. Volcanoes are found in the Cascade Range, the Hawaiian Islands, and the Aleutian Islands of Alaska. Plateaus are high flat areas. The Colorado Plateau is in the middle of the southern Rockies. Valleys are low places where rivers have worn away the rock. Canyons are valleys with steep sides. The Grand Canyon is cut by the Colorado River as it flows through the Colorado Plateau. Compared with mountains, moraines are relatively low series of hills and ridges formed from the rocks (large and small), gravel, and sand left by glaciers as they advanced and receded. There are many moraines around the Great Lakes, which were also formed by glaciers. [CC 1: Patterns]

✪ Figure 5.4. Highway Cut Through Rock

Note: A full-color version of this figure is available on the book's Extras page at *www.nsta.org/pbl-earth-space*.

Model response for grades 6–8 extension: Mountains are places where the surface of Earth has been crumpled and piled up by colliding continental plates. Where humans make cuts through mountains (for example, to build a highway), you can often see that the rocks have been folded or broken. [See Figure 5.4.] Canyons occur where the river is cutting quickly and there aren't forces like lots of rain to make the sides crumble. Mesas and buttes are formations of hard rock left after softer rock has been eroded away. [CC 2: Cause and Effect: Mechanism and Explanation]

Activity Guide

Students can find answers to this problem by looking up the word *landforms*. However, this is not the goal of the performance expectation. Students should use topographic and relief maps (see Page 3: Resources) and highway cutaways to identify the following patterns (CC 1: Patterns):

- Mountains appear as areas with high points and much irregular variation in altitude.

- Plateaus are relatively flat, but they are higher than some of the surrounding land.

- Valleys and canyons are low places with a branching structure and usually water in the lowest part.

- Moraines are very low lines of hills and ridges found only in places that were at one time glaciated. When humans cut into moraines, they are often found to be made of rocks, gravel of different sizes, sand, and clay.

To help your students understand what topographic maps are, you can have them build a model of an irregularly shaped mountain out of modeling clay. They (or you) can cut the mountain into horizontal slices of equal thickness. Students can put each slice on paper and trace around the base. They can then cut out the shapes and stack them to make a topographic map of their mountain.

Alternatively, students can use a topographic map to construct a model of the landscape. This is essentially how the fictional Cosmic Comic Book Company would have a computer generate background landscapes. Make multiple copies of a topographic map. Have each pair or group of students cut out a different contour line. They should then trace around their shape onto cardboard, foam core, or a layer of clay and cut out the shape. When the whole class's contours are stacked in the proper order, they will form a 3-D model of the landscape shown on the map. If students take top-down and side-view pictures of their models, they will be able to see the relationship between topographic maps and 3-D landscapes.

SAFETY PRECAUTIONS

1. If using sharp instruments to cut clay models, handle with care to prevent cutting or puncturing skin.

2. Review safety data sheet (SDS) for clay. Clean up clay before it dries to prevent respiratory exposure.

3. Immediately wipe up any water spilled on the floor to avoid a slip or fall hazard.

4. Wash hands with soap and water when the activity is complete.

PROBLEM 1

Assessment

An Eagle's View

Transfer Tasks

TRANSFER TASK 1

[*Note to teacher:* Give students a relief or topographic map of anywhere in the world with two places marked that were not used in other activities. The two locations should have different topography.]

Look at the map and use the patterns you've found to identify the landforms.

Model response: Areas with steep slopes have contour lines close together, while flatter areas have contour lines much farther apart. The contour lines of a rounded hill form circles, while valleys or ridges have sections of parallel lines. Mountains appear as areas with high points and much irregular variation in altitude. Plateaus are relatively flat, but higher than some of the surrounding land. Valleys and canyons are low places with a branching structure and usually water in the lowest part. Moraines are lower hills or ridges found only in places that were at one time glaciated. [SEP 2: Developing and Using Models; CC 1: Patterns]

TRANSFER TASK 2 (FOR GRADES 6–8)

Pick a place where you would like to live. Using a relief map or a topographic map as a guide, describe the landforms you would find there. Explain how and why they formed.

Model response: I would like to live in Dutton, Arkansas, because my grandparents used to live there. There are low mountains or tall hills there. Mountains form when two continental plates collide, so that must have happened a long time ago. I think the mountains aren't very tall because they have been eroding away since they stopped building up. There are lots of valleys in the mountains. Each one has a stream or river in it, at least when it rains. The streams and rivers eroded the valleys.

Application Question

Write an Eli Eagle adventure. In the adventure, describe what Eli sees as he flies from place to place. Include at least three different landforms in the adventure.

Model response: The Sun was coming up as Eli Eagle flew over the Smokey Mountains. The usual bluish haze hung over the mountains. He couldn't see much in the way of animal life because of the tree cover. But every now and then there was a river running down a valley or a farmer's field where he could see the ground. The Sun disappeared behind clouds, and it began to rain. It rained harder and harder. This didn't bother Eli, but he was on the lookout for animals in trouble. At the base of a bluff, Eli saw a bear cub trying to shelter from the rain under a small overhanging rock. The stream just below him was starting to fill up with fast-running water. From his vantage point high up in the sky, Eli could see that all of the streams were rising fast. The bear cub was trapped between the bluff and the stream! Eli swooped down and grabbed her and carried her to the other side of the mountain, where he set her down on high, dry ground. But he couldn't stay to chat. He had to check to see if all of the water running down from the mountain streams was endangering any unsuspecting animals on the plains west of the mountains. [SEP 8: Obtaining, Evaluating, and Communicating Information]

Problem 2: Diving With Dolphino

Alignment With the *NGSS*

PERFORMANCE EXPECTATIONS	• *4-ESS2-2:* Analyze and interpret data from maps to describe patterns of Earth's features. • *MS-ESS2-3:* Analyze and interpret data on the distribution of fossils and rocks, continental shapes, and seafloor structures to provide evidence of the past plate motions. • *MS-ESS2-4:* Develop a model to describe the cycling of water through Earth's systems driven by energy from the Sun and the force of gravity.
SCIENCE AND ENGINEERING PRACTICES	• Developing and Using Models • Analyzing and Interpreting Data • Constructing Explanations and Designing Solutions • Obtaining, Evaluating, and Communicating Information

DISCIPLINARY CORE IDEAS	• *ESS1.C: The History of Planet Earth*
	
	  ○ *Grades 3–5:* Local, regional, and global patterns of rock formations reveal changes over time due to earth forces, such as earthquakes. The presence and location of certain fossil types indicate the order in which rock layers were formed.
	
	  ○ *Grades 6–8:* Tectonic processes continually generate new ocean seafloor at ridges and destroy old seafloor at trenches.
	
	• *ESS2.B: Plate Tectonics and Large-Scale System Interactions*
	
	  ○ *Grades K–2:* Maps show where things are located. One can map the shapes and kinds of land and water in any area.
	
	  ○ *Grades 3–5:* The locations of mountain ranges, deep ocean trenches, ocean floor structures, earthquakes, and volcanoes occur in patterns. Most earthquakes and volcanoes occur in bands that are often along the boundaries between continents and oceans. Major mountain chains form inside continents or near their edges. Maps can help locate the different land and water features of Earth.
	
	  ○ *Grades 6–8:* Maps of ancient land and water patterns, based on investigations of rocks and fossils, make clear how Earth's plates have moved great distances, collided, and spread apart.

DISCIPLINARY CORE IDEAS (*continued*)	• *ESS2.C: The Roles of Water in Earth's Surface Processes*
	
	  ○ *Grades K–2:* Water is found in the ocean, rivers, lakes, and ponds. Water exists as solid ice and in liquid form.
	
	  ○ *Grades 6–8:* Water continually cycles among land, ocean, and atmosphere via transpiration, evaporation, condensation and crystallization, and precipitation, as well as downhill flows on land.

CROSSCUTTING CONCEPTS	• Patterns • Cause and Effect: Mechanism and Explanation • Systems and System Models

Keywords and Concepts

Landforms, geologic forces, plate tectonics

Problem Overview

Students help a comic book publisher develop scenes and images of landforms found below the surface of the ocean.

Images available in full color on the Extras page (*www.nsta.org/pbl-earth-space*) are marked with the following icon: ✪.

PROBLEM 2

Page 1: The Story

Diving With Dolphino

The Cosmic Comic Book Company has just hired your class. The company is planning several new superheroes. One of these is Dolphino, a silver dolphin (see Figure 5.5) who can swim for long stretches going 300 miles per hour. She can "sprint" for short distances at 500 miles per hour.

The comic book company will use a computer and real maps to create the background pictures. They have read that oceans have very exciting landforms. For example, on land in the United States, the tallest mountain is Mount Denali (formerly called Mount McKinley) in Alaska (20,310 feet [6,190 meters]) and the second tallest mountain is Mount Whitney in California (14,505 ft. [4,421 m]). But there are tall mountains in the oceans, too. Mauna Kea (Hawaii) is considered to be the highest mountain on Earth because it rises from the ocean floor 33,500 ft. (10,211 m).

✪ **Figure 5.5. Dolphin**

Many earthquakes occur under the ocean. There are also many volcanic eruptions. These will make exciting places for Dolphino to visit. But the big, flat areas aren't as interesting. The writers and animators (people who create the pictures) need your help so that they can make exciting adventures for Dolphino. What types of landforms are there in the ocean, and where are they likely to find them?

Your Challenge:

- *Identify the patterns of landforms on the ocean floors. Explain how they compare to landforms on land.*

- **Grades 6–8 extension:** *Find out how and why earthquakes and eruptions are associated with particular landforms.*

PROBLEM 2

Page 2: More Information

Diving With Dolphino

One of the comic book writers came across the following interesting facts that might be useful when planning the scenes for Dolphino's adventures:

- The ocean off the coast of North Carolina is quite shallow (less than 500 ft. [152 m] deep). However, after you go about 50 miles (80 kilometers) from shore, it quickly gets much deeper (more than a mile and a half deep).

- The deepest point in the ocean is the Mariana Trench. It is off the coast of Guam. It is roughly 7 miles (13 km) deep.

- The Ring of Fire is a region with a string of active volcanoes and earthquake zones that ring the Pacific Ocean (see Figure 5.6).

Your Challenge:

- *Identify the patterns of landforms on the ocean floors.*

- *Explain how they compare to continental landforms.*

- **Grades 6–8 extension:** *Explain how and why earthquakes and volcanic eruptions are associated with particular landforms.*

✪ **Figure 5.6. Underwater Volcano**

PROBLEM 2

Page 3: Resources

Diving With Dolphino

- Relief maps of the ocean floor—see, for example, *www.orangesmile.com/ travelguide/afoto/ocean-maps.htm,* or search online images for *relief maps of ocean floor*

- Maps of volcanic eruptions and earthquakes—see, for example, *www.pbs.org/ wgbh/nova/education/activities/2515_vesuvius.html* and *http://hvo.wr.usgs.gov/seismic/ volcweb/earthquakes,* or search online images for *map of volcanoes and earthquakes*

- Maps of tectonic plates—see, for example, *https://planetjan.wordpress. com/2010/04/04/its-the-end-of-the-world-as-we-know-it/map_plate_tectonics_world,* or search online images for *map of tectonic plates*

- A scaled infographic comparing the heights and depths of terrestrial and oceanic landforms, available at *www.livescience.com/29536-infographic-tallest- mountain-to-deepest-ocean-trench.html*

PROBLEM 2

Teacher Guide

Diving With Dolphino

Problem Context

Students use topographic maps to find patterns in oceanic landforms. The landforms are not difficult to see on the maps. However, if you feel that your students might be overwhelmed by looking at a world map, you can have them start by looking at the Pacific Ocean, which has most of the main features.

Problem Solution

Model response for grades 3–5: Continental shelves are relatively shallow, flat areas off the coast of some continents. Trenches are another feature associated with some coastlines; trenches are narrow, but often very deep, "valleys." In the middle of the oceans and extending to some continents are oceanic ridges. Each oceanic ridge is evident as a high and very long ridge crisscrossed with perpendicular fault lines. The oceanic ridges are not as wide as most terrestrial mountain ranges. The Pacific and Atlantic mid-oceanic ridges are much longer than any terrestrial mountain range. Mountain ranges are visible on some parts of the ocean floor as well as on land. [CC 1: Patterns]

Model response for grades 6–8 extension: Oceanic trenches are much deeper than terrestrial valleys and are caused by tectonic activity (subduction). They occur where a denser oceanic plate is subducted under a lighter continental plate. Mid-oceanic ridges are also caused by tectonic activity (diverging plates). Where two plates diverge, the crust is torn apart and lava wells up, forming ridges along the tear. Other manifestations of the tectonic activity associated with trenches and ridges are earthquakes and volcanic eruptions. [CC 2: Cause and Effect: Mechanism and Explanation]

Activity Guide

If your class did not do Problem 1: An Eagle's View and you would like your students to understand topographic maps with contour lines, you can do the hands-on activities suggested in the activity guide for that problem (see p. 76).

As in Problem 1, students could answer this problem by looking up a term—in this case, *oceanic landforms*. However, as in Problem 1, this is not the goal of the performance expectation. Students should use relief maps (see Page 3: Resources) to identify patterns in features (SEP 4: Analyzing and Interpreting Data). They will spot the oceanic ridges with their characteristic long broad ridges, with many fault lines running perpendicular to them. Various mountain ranges are also evident. Continental shelves and subduction zone trenches are harder to see. The information on Page 2 should direct them to look more closely at these. Once students have identified the main features, they can look at a map of earthquake and volcanic activity to see if any landforms are associated with these events.

Middle school students can go a step further with this problem to see how landforms map onto tectonic plates. They will find that the arrows on maps of the tectonic plates indicate the direction that the plates are moving. Converging oceanic and continental plates are associated with trenches and mountain ranges containing volcanoes. Converging continental plates are associated with active mountain-building sites such as the Himalayas. Diverging oceanic plates are where ridges form. [CC 1: Patterns; CC 2: Cause and Effect: Mechanism and Explanation]

PROBLEM 2

Assessment

Diving With Dolphino

Transfer Tasks

TRANSFER TASK 1

[*Note to teacher:* Give students a relief or topographic map of the ocean floor with a particular feature marked.]

Look at the map and draw a picture of what the marked feature would look like as seen from a submarine or underwater camera. Make a caption for your picture that includes the type of feature and a description of where such features are typically found.

> *Model response:* If you could travel by submarine around Australia, you would see continental shelves. These are relatively shallow, flat areas off the coast. In the southwest, if you travel away from the continent, the ocean floor drops away very quickly. The ocean gets deep fast. In the northeast, the continental shelf is pretty big. But if you travel far enough, you will come to mountains on the ocean floor. [CC 1: Patterns]

TRANSFER TASK 2 (FOR GRADES 6–8)

[*Note to teacher:* The question below is only appropriate for students who identified the relationship between the oceanic (and terrestrial) landforms and tectonic plates and their movement (see the "Activity Guide" section in this problem's teacher guide.) It asks students to construct explanations (SEP 6: Constructing Explanations and Designing Solutions) by using a systems approach (CC 4: Systems and System Models).]

Which type of plate boundary do you think is safest for Dolphino: convergent oceanic plates, convergent oceanic and continental plates, or divergent oceanic plates? Explain your answer.

> *Model response:* All of these boundaries are associated with earthquakes. Volcanoes form when oceanic plates subduct under continental plates and at mid-oceanic rifts. So maybe convergent oceanic plates are safest, since there are fewer volcanoes. [CC 2: Cause and Effect: Mechanism and Explanation]

CHAPTER 5

TRANSFER TASK 3 (FOR GRADES 6–8)

The Rocky Mountains are a major feature of the western United States. They formed more than 50 million years ago and are no longer building up. Use what you know about plate tectonics to explain how they formed and what is happening to them now.

> *Model response:* Mountains form where plates collide, so oceanic or continental plates must have converged on the North American plate and merged with it. [CC 2: Cause and Effect: Mechanism and Explanation] Now that it's all one plate, no mountain building is going on and the Rockies are slowly being eroded away. [SEP 6: Constructing Explanations and Designing Solutions]

TRANSFER TASK 4 (FOR GRADES 9–12)

Fossils of marine animals and sandstone formed from beaches have been found high in the Rocky Mountains. In some places, the sandstone shows ripple marks from waves and dinosaur footprints. Explain how sea fossils and sandstone could end up thousands of feet above sea level and exposed for people to see.

> *Model response:* The land the Rockies sit on now used to be the western edge of the North American continent. Continued collisions with other plates added new terranes (fragments of crust that breaks off of one plate and attaches to another) to western North America and deformed, buckled, and crumpled up the flat, sandy beaches and remains of marine animals that were the original environment. [CC 2: Cause and Effect: Mechanism and Explanation] That is why you find ripple marks with dinosaur tracks on the sides of the Rockies. The fossils formed when that land was at the edge of the ocean. Later that plate collided with another continental plate, breaking it into pieces that piled up. Once they were exposed to weather, the pieces were slowly eroded, revealing the fossils. [SEP 6: Constructing Explanations and Designing Solutions]

TRANSFER TASK 5 (FOR GRADES 9–12)

The oceanic ridges are active places where the ocean is spreading and new ocean crust is forming. Does this mean that Earth is getting bigger? Explain your answer.

> *Model response:* No, Earth remains the same size because at the same time that crust is forming at divergent boundaries such as the oceanic ridges, crust is being subducted and recycled at convergent boundaries.

Application Question

Write a story for one of Dolphino's adventures. Pick a real underwater location and include ocean landforms that she might encounter there.

> *Model response:* [*Note to teacher:* Students' responses will depend on the location they choose, but should accurately reflect the features of that area.] I am Dolphino, oceanic superhero! Let me tell you about my latest adventure. It was Tuesday and I was patrolling the Atlantic between Canada and Europe. I usually make a clockwise circuit. I like to cruise above the continental shelf off the coast of Canada where there are lots of fish. Then I headed east across the Atlantic. I was adjusting my depth so that I could hear the whale songs better, when all of a sudden there was a loud rumble and I was being scalded by hot water in the dark. I headed up to try to get away from it. The usual currents were disrupted and I was getting lost. There were more rumbles and I decided I better get out of there. I was probably over the Mid-Atlantic Ridge. There aren't many fish there that would need my help anyway. An hour later, I felt the water get shallower. I was over the continental shelf of Britain. I was farther north than usual, but it was easy to get back on course. Nothing stops Dolphino! [SEP 8: Obtaining, Evaluating, and Communicating Information]

Problem 3: Water, Water Everywhere

Alignment With the *NGSS*

PERFORMANCE EXPECTATIONS	• *2-ESS2-3:* Obtain information to identify where water is found on Earth and that it can be solid or liquid. • *5-ESS2-2:* Describe and graph the amounts and percentages of water and freshwater in various reservoirs to provide evidence about the distribution of water on Earth. • *MS-ESS2-1:* Develop a model to describe the cycling of Earth's materials and the flow of energy that drives this process. • *MS-ESS2-4:* Develop a model to describe the cycling of water through Earth's systems driven by energy from the Sun and the force of gravity.
SCIENCE AND ENGINEERING PRACTICES	• Developing and Using Models • Analyzing and Interpreting Data • Constructing Explanations and Designing Solutions • Obtaining, Evaluating, and Communicating Information

DISCIPLINARY CORE IDEAS	• *ESS2.A: Earth Materials and Systems*
	
	  ○ *Grades K–2:* Wind and water can change the shape of the land.
	
	  ○ *Grades 3–5:* Rainfall helps to shape the land and affects the types of living things found in a region. Water, ice, wind, living organisms, and gravity break rocks, soils, and sediments into smaller particles and move them around.
	
	  ○ *Grades 6–8:* All Earth processes are the result of energy flowing and matter cycling within and among the planet's systems. This energy is derived from the Sun and Earth's hot interior. The energy that flows and matter that cycles produce chemical and physical changes in Earth's materials and living organisms.
	
	• *ESS2.C: The Roles of Water in Earth's Surface Processes*
	
	  ○ *Grades K–2:* Water is found in the ocean, rivers, lakes, and ponds. Water exists as solid ice and in liquid form.
	
	  ○ *Grades 3–5:* Nearly all of Earth's available water is in the ocean. Most freshwater is in glaciers or underground; only a tiny fraction is in streams, lakes, wetlands, and the atmosphere.
	
	  ○ *Grades 6–8:* Water continually cycles among land, ocean, and atmosphere via transpiration, evaporation, condensation and crystallization, and precipitation, as well as downhill flows on land. Global movements of water and its changes in form are propelled by sunlight and gravity.

CROSSCUTTING CONCEPTS	• Patterns
	• Cause and Effect: Mechanism and Explanation
	• Systems and System Models
	• Energy and Matter: Flows, Cycles, and Conservation

Keywords and Concepts

Water cycle, erosion, gravity, heating and cooling

Problem Overview

Students help a comic book publisher understand the path of a drop of water traveling through the water cycle.

Images available in full color on the Extras page (*www.nsta.org/pbl-earth-space*) are marked with the following icon: ○.

PROBLEM 3

Page 1: The Story

Water, Water Everywhere

Your class has just been hired by the Cosmic Comic Book Company. The company is planning several new superheroes. One of these is Dropedo, who does not look like a typical superhero. She is tiny—the size of a drop of water—and is clear like water—you can see through her. And she can change shape and go wherever liquid water can go. She is happy in freshwater and saltwater, but her color changes depending on whether she is in freshwater or saltwater.

For two other superheroes, the comic book company is using maps and computers to create the background pictures. But the writers are not sure how they will create the backgrounds for Dropedo's adventures. Topographic maps will tell them where she can flow downhill in rivers that lead to lakes, larger rivers, or the ocean (see Figure 5.7). On the maps, rivers and other bodies of water are usually shown in blue. But some water soaks into the ground, and groundwater is not shown on the maps. Water comes out of the ground in some places such as springs and is connected to rivers and streams. Maps also don't show that water goes in and out of living things. So before the writers can write Dropedo's adventures, they need your class to help them learn more about Earth's water than what they can see on maps.

✪ Figure 5.7. Ocean and River

Your Challenge: *Answer the writers' questions:*

- *Where will Dropedo be in freshwater? Where will she be in saltwater?*
- *How much of Earth's water is fresh and how much is salty?*
- *Where are all of the places water can go?*
- **Grades 6–8 extension:**
  - *What are the connections between all of the places water can go?*
  - *How powerful is water? What makes water move and change?*

NATIONAL SCIENCE TEACHERS ASSOCIATION

PROBLEM 3

Page 2: More Information

Water, Water Everywhere

Dropedo is tiny and can go wherever liquid water goes. The writers think that this will lead to some interesting adventures. They would like to know more about how water travels through rock cracks, soil, and living things (see Figures 5.8–5.10).

✪ Figure 5.8. Plant Xylem (magnified photo)

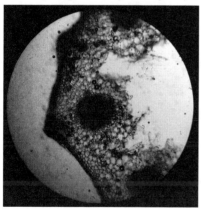

✪ Figure 5.9. Cracks in Rock (magnified photo)

Note: Full-color versions of these figures are available on the book's Extras page at *www.nsta.org/pbl-earth-space*.

✪ Figure 5.10. Soil in Flowerpot (magnified photo)

The writers have also heard that water is powerful, perfect for a superhero. But they need to know more about this.

Your Challenge: *Answer the writers' questions:*

- *Where will Dropedo be in freshwater? Where will she be in saltwater?*
- *How much of Earth's water is fresh and how much is salty?*
- *Where are all of the places water can go?*
- **Grades 6–8 extension:**
  - *What are the connections between all of the places water can go?*
  - *How powerful is water? What makes water move and change?*

PROBLEM 3

Page 3: Resources

Water, Water Everywhere

- Information on the amount of water on Earth: *http://water.usgs.gov/edu/gallery/global-water-volume.html*.

- Holling, H. C. 1980. *Paddle-to-the-sea*. Boston: HMH Books for Young Readers; Mason, B. 2015. *Paddle to the sea* [film adaptation of the book by Holling]. National Film Board of Canada. *www.nfb.ca/film/paddle_to_the_sea*.

- Harvard-Smithsonian Center for Astrophysics. 2007. *The habitable planet: A systems approach to environmental science*. Unit 8: Water resources. Annenberg Learner [video series]. Los Angeles: Annenberg Foundation. *www.learner.org/resources/series209.html#program_descriptions*.

- Relf, P. 1996. *The magic school bus wet all over: A book about the water cycle*. New York: Scholastic; The Magic School Bus video series available at *www.scholastic.com/magicschoolbus/tv/index.htm*.

PROBLEM 3

Teacher Guide

Water, Water Everywhere

Problem Context

This problem focuses on where water can be found on Earth and how much of it is fresh versus salty. Students draw on maps, personal experience, and videos. Local wetlands, water treatment plants, or other water features can also be used to help students identify where water can be found.

Problem Solution

Model response for grades 3–5: Water exists in all three states of matter on Earth. It is found in many places, but not all places have the same amount of water. Liquid water can be found on the surface of Earth, underground, and in the atmosphere as clouds or as rain. Solid water (ice) is found in glaciers and polar ice caps and in the atmosphere as snow or hail, or ice crystals in clouds. Water vapor is found in the atmosphere. Liquid water can be either fresh (groundwater, lakes, rivers) or salty (oceans). It is visible as surface water, but it also exists underground. Surface water and groundwater are connected. When it rains, some of the water runs over the surface of the ground. It moves downhill and ends up in streams and rivers. Some of the water has time to soak into the ground. It, too, moves down and may find its way into underground reservoirs or it may reconnect to surface water as it travels through open spaces in soil and rock layers. [CC 5: Energy and Matter: Flows, Cycles, and Conservation]

Almost all the water (~97%) on Earth is found in the oceans. The second largest source (~2%) of water is ice. Less than 1% of the water on Earth is freshwater in the liquid state, and it is not evenly distributed over Earth's surface. [SEP 5: Using Mathematics and Computational Thinking]

Dropedo will be able to go all over Earth's surface, including

- down through tiny cracks in rocks,
- down through soil and sand,
- into plants through their roots,
- out of plants through their leaves,

- into animals as part of their food or with the water they drink,

- out of animals in their breath or excrement, and

- downhill with surface water into river systems and eventually the ocean.

Model response for grades 6–8 extension: Gravity moves water downhill. Regular water can't go uphill and Dropedo cannot evaporate, so the only way she can go uphill is with her superpowers. Regular water goes up through evaporation. Plants can move water uphill. They take up liquid water from the soil and return some to the atmosphere as water vapor through transpiration. All of the other processes for moving water up into the atmosphere require heat directly or indirectly from the Sun. Heat also determines what state water is in—solid, liquid, or gas. Heat needs to be added to melt ice to form a liquid and evaporate liquid forming a gas. Heat is released when water vapor condenses to form liquid or liquid freezes to form ice.

If Dropedo is as powerful as water, she will be able to carry things (buildings, cars, rocks) and soil and sand downstream. She can also wear down rocks. [See Figure 5.11 for images of the power of water.]

❂ Figure 5.11. The Power of Water

(a) Flood

(b) Aerial view of river delta

(c) Eroded gravestone

Activity Guide

The resources on Page 3 are meant to seed students' ideas of where water can be found. These resources are particularly important for students with little experience with natural water features. These resources can be augmented by classroom models of the water cycle where students build a clay mountain in a container containing about an inch of water representing the ocean. They can pour water onto the mountain, simulating rain, and put a clear cover on the container. If they shine a bright light on the part of the water and put a petri dish of ice over the mountain, they will see condensation form near the ice, indicating that water evaporated from the "ocean" and condensed in the "clouds" above the mountain.

Students can experience the power of erosion through any of several activities. There are often signs (small gullies in the soil) of water erosion in school yards where water drips off a roof or runs out of a downspout. Students can collect water samples from local streams and rivers after a hard rain when the water is visibly "muddy." They can filter the water using a coffee filter to determine how much sediment was in it. They can model the effects of waves by making a model mountain of moist sand or soil in one end of a baking dish. If they put between a ½ inch and 1 inch of water (about 1.3–2.5 cm) in the baking dish and then gently move the end of the baking dish away from the mountain in an up-and-down motion, they will create waves that will lap at the mountain, eventually destroying it.

Students can make a more elaborate and quantitative model of soil erosion using a plastic sled filled with local soil. They should raise the sled at a slight angle with a catch basin under the lower end. Using a watering can to water the upper edge of the soil will simulate rain. Students can filter and weigh the sediment in the catch basin after each time they empty the watering can. They can study the effects of having the sled at different angles and changing the rate of rain.

Students can investigate the permeability of different soils by making drainage columns from large paper cups. They should poke a small hole in each cup using a pencil and try to keep the hole sizes equal (you may choose to do this ahead of time). They should put a disk of coffee filter or paper towel in the bottom of each cup and mark the side of the cup about 3 inches (7.6 cm) up. They can fill each cup to the line with different soils (having different amounts of organic matter, sand, clay, etc.), measure the permeability by adding 100 milliliters of water to each cup while holding it over a catch basin, and time how long it takes for the water to run through.

SAFETY PRECAUTIONS

1. Indirectly vented chemical-splash goggles, aprons, and nonlatex gloves are recommended for these activities.

2. Review safety data sheet (SDS) for clay. Clean up clay before it dries to prevent respiratory exposure.

3. Immediately wipe up any water spilled on the floor to avoid a slip or fall hazard.

4. Wash hands with soap and water when the activity is complete.

PROBLEM 3

Assessment

Water, Water Everywhere

Transfer Tasks

TRANSFER TASK 1

Write a story about the superhero Dropedo falling in your school yard. Describe what happens to her and what she "sees." You should include three different places that she passes through and at least two different states (solid, liquid, gas). [SEP 8: Obtaining, Evaluating, and Communicating Information; CC 5: Energy and Matter: Flows, Cycles, and Conservation]

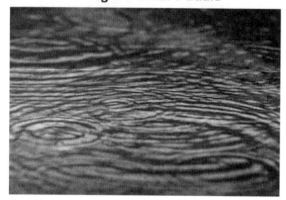

✪ **Figure 5.12. Puddle**

Model response: Ouch! I just landed on the playground pavement. Lots of other drops here, too, so we made a puddle. [Figure 5.12] Oh look, we're moving along in this spot that's slightly lower. Bump! I hit the grass. Ahh, this is comfy. I'm drifting down through the soil. Lots of little holes for me here. I see lots of grains of sand, black soil. Hi, Mrs. Earthworm. My, aren't these grass roots long. Hmm, clay. It's going to take me longer to get through this. Yes, things are going slowly now. The holes are a lot smaller. Seems like we've been drifting down for days. I might have to use my superpowers to get out of here. Oh, hello, groundwater. Lots of water here, though no one seems to be moving very much. What's that I hear? A river! And I'm off to a new place. Hmmm, seems to be a lake. My buddies on the surface are evaporating. I think I'll join them in the air. Great view from up here. [SEP 8: Obtaining, Evaluating, and Communicating Information; CC 5: Energy and Matter: Flows, Cycles, and Conservation]

TRANSFER TASK 2 (FOR GRADES 6–8)

The Grand Canyon was made by the Colorado River as it flowed over the Colorado Plateau. Explain how this happened. [SEP 6: Constructing Explanations and Designing Solutions; CC 2: Cause and Effect: Mechanism and Explanation]

Model response: Canyons are formed when a river runs fast through fairly hard rock. [CC 2: Cause and Effect: Mechanism and Explanation] This means the Colorado Plateau must be sloped. As the Colorado River ran quickly along its path, it eroded away the rock below it, carrying the rock bits along with it. The sides of the canyon didn't crumble very much, leaving nearly vertical walls in some places. [SEP 6: Constructing Explanations and Designing Solutions]

Application Question

Describe two different ways (one involving the atmosphere and the other *not* involving the atmosphere) that rainwater falling near you could end up in the ocean.

Model response: The rain could evaporate off the ground into the atmosphere. Once the water vapor is in the air, if that air mass moves out over the ocean, the water could condense and fall in the ocean. A different way for the rain to get to the ocean would be for it to run downhill into whatever streams are nearby. Eventually the streams empty into rivers and the rivers flow to the ocean. The rain might also seep into the ground and eventually end up in streams through the connection between groundwater and surface water. [SEP 6: Constructing Explanations and Designing Solutions; CC 1: Patterns]

Earth's Landforms and Water Problems: General Assessment

General Questions

GENERAL QUESTION 1 (FOR ELEMENTARY GRADES)

Describe several types of landforms you might find if you compare different regions around the world.

Model response: [*Note to teacher:* Students may not describe all the potential landforms and forces, but the model response illustrates some examples of the types of responses you may expect to see.] Common landforms include mountains, volcanoes, and plateaus. Mountains are areas with high, steep areas that are higher and steeper than hills. Volcanoes can also build mountains. Plateaus are relatively flat areas that are higher than surrounding areas. They usually occur between two mountain ranges, like the plateau between the Rockies and the Cascade Range. [CC 1: Patterns]

Rivers erode valleys into the landscape. Deep valleys with steep sides we call canyons. The Grand Canyon and the New River Gorge (in West Virginia) are examples. Glaciers can create landforms, too. When they advance, they flatten out the bottom of the valleys that they move through by pushing soil and rock along the surface. Where they stop for some time (hundreds to thousands of years), long ridges of rock and soil are left behind in ridges called moraines. Digging into these moraines usually shows a mix of different types of rock, gravel, and soils. The glacier can also leave behind glacial lakes ("kettle lakes") in the low areas that are left filled with melting ice. The lakes of Minnesota, Michigan, Wisconsin, and northern Indiana are good examples of these lakes. [CC 2: Cause and Effect: Mechanism and Explanation]

GENERAL QUESTION 2 (FOR MIDDLE SCHOOL EXTENSION)

Explain what kinds of natural forces can create these landforms.

Model response: [*Note to teacher:* Students may not describe all the potential landforms and forces, but the model response illustrates some examples of the types of responses you may expect to see.] When tectonic plates collide, Earth's crust crumbles and piles up, creating high points that we call mountains. Volcanoes are also the product of tectonic activity. They can occur in two ways. Either the crust is thin and magma floats to the surface at "hot spots" as in Hawaii or an oceanic plate subducts under a continental plate. There are a lot of volcanoes around the edge of the Pacific Ocean in an area called the Ring of Fire.

Where there is no tectonic activity, erosion is the predominant force changing Earth's surface. Water, under the force of gravity, moves downhill, eroding valleys. Under special conditions in some areas, deep canyons form where, over long periods of time, a river cuts into harder rocks and there aren't other forces making the sides of the valley crumble.

Water can also create mesas and buttes where there are smaller areas of hard rock in soft rock. Water (and wind) erodes the softer rocks, leaving the harder rock as a small flat-topped mesa or butte jutting out. [CC 2: Cause and Effect: Mechanism and Explanation]

Common Beliefs

Indicate whether the following statements are true (T) or false (F), and explain why you think so *(model responses shown in italics).*

1. The bottom of the ocean is flat. *(F) Although areas of the ocean floor are flat, mountains, ridges, and very deep valleys can be found on the ocean floor.*

2. Dripping water can erode a rock. *(T) Over long periods of time, the dripping water carries away the rock. The water may dissolve the rock or break off tiny bits.*

3. Submarines can go under islands. *(F) Islands are mountains rising from the ocean floor to the surface. There is no ocean below them.*

4. (For grades 6–8) Today's mountain ranges have always been on the surface of Earth, but their elevations/heights change over time. *(F) Today's mountain ranges formed at different times when two tectonic plates collided. Some (like the Himalayas) are still being piled up. Others (like the Rockies and the Appalachians) are no longer actively building. They are gradually being worn down by weathering and erosion.*

5. (For grades 6–8) Rainwater that falls on soil and soaks into the ground may end up in a nearby stream. *(T) The water may seep through soil and rocks into the stream. The many holes and gaps are connected.*

ROCK CYCLE AND PLATE TECTONICS

T he problems in this chapter explore the ways in which characteristics of rocks tell a story about their formation and changes over time. The rock cycle and plate tectonics are key concepts addressed in the chapter.

The chapter emphasizes the crosscutting concepts (CCs) described in the *Next Generation Science Standards* (*NGSS*; NGSS Lead States 2013), especially Patterns (CC 1); Cause and Effect: Mechanism and Explanation (CC 2); Systems and System Models (CC 4); and Energy and Matter: Flows, Cycles, and Conservation (CC 5). The problems also help students develop the scientific practices of Developing and Using Models (*NGSS* science and engineering practice [SEP] 2); Analyzing and Interpreting Data (SEP 4); Constructing Explanations and Designing Solutions (SEP 6); and Obtaining, Evaluating, and Communicating Information (SEP 8). The disciplinary core ideas (DCIs) addressed in this chapter are Earth Materials and Systems (ESS2.A) and Plate Tectonics and Large-Scale System Interactions (ESS2.B).

Big Ideas
Rock Cycle

- Igneous, metamorphic, and sedimentary rocks are indicators of geologic and environmental conditions and processes that existed in the past. These processes include melting, cooling and crystallization, weathering and erosion, sedimentation and lithification, and metamorphism. In some way, all of these processes are influenced by plate tectonics, and it can be said that plate tectonics drives the rock cycle.

Plate Tectonics

- Earth's lithosphere is broken into large mobile pieces called tectonic plates that float on the mantle. Tectonic plates are characterized by the kind of rock they contain. In general, *oceanic plates* are denser than continental plates and are made of dark-color rocks, mostly basalts. *Continental plates* are less dense and are made of lighter-color rocks such as quartz and granites. However, some plates, such as the North American Plate, are a mixture of continental and oceanic plates that have come together over time.

- Oceanic plates are created at mid-oceanic ridges by magmatic activity. The outer edges of the growing oceanic plate may collide with another oceanic plate or with a continental plate. The denser oceanic plate will be subducted under the plate (lighter oceanic or continental), thus forming, respectively, a trench or a trench and mountains interspersed with volcanoes. The other type of *convergent boundary* (where plates are colliding) occurs when two continental plates collide. As they are pushed together, the two light plates crumble and pile up, forming a thick region in the plates that is a mountain range.

- Places where plates are separating are called *divergent margins. Convergent margins* are where plates come together and where subduction occurs. *Transform boundaries* are boundaries along which plates slide by each other.

- Plate boundaries can change over time. For example, the San Andreas Fault has developed along a former subduction zone. The Keweenaw Peninsula went from *extension* (plates spreading apart) to *compression* (plates pressing against each other).

Rock Characteristics as Evidence of the Past

- Earth has a long (4.6 billion years) geologic history that can be understood by studying the rocks. Earth processes we see today are similar to those that occurred in the past. Spatial relationships and ages of rocks can tell us about the large-scale tectonic history of a region.

- Rocks tell us about their origin and history through texture (size and shape of crystals) and composition. In terms of texture of igneous rocks, large crystals indicate that a rock cooled slowly underground. Lack of crystals indicates rapid cooling above ground.

- Sedimentary rocks can form in the low places around convergent boundaries when uplifted rocks are broken down by wind and water, creating sediments. Sedimentary rocks containing larger pieces of recycled rock formed near the source of those rocks, often near mountains or higher ground. Smaller pieces of stable sediments such as quartz indicate that the sediment has traveled farther from its source or has experienced extreme weathering.

- Volcanic rocks at subduction zones are different from those at rifts. At subduction zones, some of the volcanic rocks are likely to come from granitic rocks from the continental plate and some from basaltic rocks from the oceanic plate. Divergent plate boundaries or rifts are usually between oceanic plates and produce basalts.

- In subduction zones, sedimentary rocks are created from the erosion of volcanoes, the uplift of the continental margin, and *accretion* (scraping off) from the oceanic plate.

- When plates converge, metamorphic rocks are created from preexisting rocks.

Conceptual Barriers

The rock cycle is difficult to visualize because of the long time span of the process and location of the events below the surface of Earth. Plate tectonics is another process that happens very slowly and involves moving objects that are very large in scale.

Common Problems in Understanding

- Geological events happen on spatial and temporal scales beyond human experience, so students have trouble understanding the following ideas:

- Deep time and change over time can be difficult to understand.

- Size and scale related to plates and plate margins can be difficult to understand.

- Most representations describe tectonic plates as being big and rigid and changing only on the margins.

- Volcanoes are typically shown only as a steep-sided composite type.

- The rock cycle is often oversimplified and represented as a strict cycle. The diagrams often do not show the multiple possible events that can change a rock type.

Common Misconceptions

- Pangaea was the original continent and represents the beginning of plate tectonics on Earth.

- A continent is a plate. The edge of a continent is a plate boundary.

- All plates are made of the same materials.

- The plate boundary type is the same as the plate itself—divergent boundary = divergent plate. The type of boundary at one edge of the plate defines the plate.

- Continents are "floating" islands, and only continents move.

- All plates move in the same direction and at the same speed.

- Plate motion is too slow to be measured—and the distance too insignificant to be noticed.

- New rocks form when "big" rocks are broken down and become little rocks—rather than through change at the molecular level.

- The asthenosphere is liquid. The lower mantle is liquid.

- Plates move but have no effect on one another.

- Molten rock flows to the surface from the middle of Earth.

- Erosion is the only process that alters Earth.

- Volcanoes and earthquakes always occur at or near plate boundaries. They cannot occur in the interior of plates.

- New rock is added to plates only from the top when volcanoes spew out molten rock that solidifies into new rock on the surface of the plate.

Interdisciplinary Connections

There are a number of interdisciplinary connections that can be made with the problems in this chapter. For example, connections can be made to literature, language arts, geography, mathematics, and art and music (see Box 6.1).

Box 6.1. Sample Interdisciplinary Connections for Rock Cycle and Plate Tectonics Problems

- **Literature:** Read books about seismic and volcanic events (e.g., *Tambora: The Eruption That Changed the World* by Gillen D'Arcy Wood) and discuss the literary techniques used to create descriptive stories.

- **Language arts:** Write a descriptive story about the forces that produced a prominent local landform or the way people might have lived before that landform was created; write a story about exploration of a deep open trench in a submarine.

- **Geography:** Use maps to identify regions around the globe with similar plate boundaries; examine the impact of tectonics on the people living in areas with seismic and volcanic activity.

- **History:** Read about and discuss important historical events that were influenced by earthquakes and volcanoes, such as the eruption of Vesuvius or the tsunami in Alaska in 1958.

- **Mathematics:** Use data about shifts along the San Andreas Fault line or the separation of North America from Europe to find rates of tectonic movement; find the volume of magma required to create the basalt flows in the Keweenaw Peninsula; measure and calculate densities of different types of rocks.

- **Art and music:** Study the use of igneous rocks for art and jewelry (e.g., obsidian or quartz) or rocks such as cinnabar, hematite, and gypsum used in art as pigments, clays, and surfaces on which artists paint or draw.

References

NGSS Lead States. 2013. *Next Generation Science Standards: For states, by states.* Washington, DC: National Academies Press. *www.nextgenscience.org/next-generation-science-standards.*

Wood, G. D. 2014. *Tambora: The eruption that changed the world.* Princeton, NJ: Princeton University Press.

Problem 1: Keweenaw Rocks

Alignment With the *NGSS*

PERFORMANCE EXPECTATIONS	• *MS-ESS2-1:* Develop a model to describe the cycling of Earth's materials and the flow of energy that drives this process. • *MS-ESS2-2:* Construct an explanation based on evidence for how geoscience processes have changed Earth's surface at varying time and spatial scales. • *HS-ESS2-1:* Develop a model to illustrate how Earth's internal and surface processes operate at different spatial and temporal scales to form continental and ocean-floor features.
SCIENCE AND ENGINEERING PRACTICES	• Developing and Using Models • Analyzing and Interpreting Data • Constructing Explanations and Designing Solutions • Obtaining, Evaluating, and Communicating Information

DISCIPLINARY CORE IDEAS	• *ESS2.A: Earth Materials and Systems*
	
	  ○ *Grades 6–8:* All Earth processes are the result of energy flowing and matter cycling within and among the planet's systems. This energy is derived from the Sun and Earth's hot interior. The energy that flows and matter that cycles produce chemical and physical changes in Earth's materials and living organisms.
	
	• *ESS2.B: Plate Tectonics and Large-Scale System Interactions:*
	
	  ○ *Grades 6–8:* Maps of ancient land and water patterns, based on investigations of rocks and fossils, make clear how Earth's plates have moved great distances, collided, and spread apart.
	
	  ○ *Grades 9–12:* Plate tectonics is the unifying theory that explains the past and current movements of the rocks at Earth's surface and provides a framework for understanding its geologic history. Plate movements are responsible for most continental and ocean-floor features and for the distribution of most rocks and minerals within Earth's crust.

CROSSCUTTING CONCEPTS	• Patterns • Cause and Effect: Mechanism and Explanation • Systems and System Models • Energy and Matter: Flows, Cycles, and Conservation

Keywords and Concepts

Rock cycle, formation of rocks, plate tectonics, divergent margin, rift zone

Problem Overview

A girl wonders how volcanic rocks and sedimentary rocks ended up in the same area of the Upper Peninsula of Michigan.

Images available in full color on the Extras page (*www.nsta.org/pbl-earth-space*) are marked with the following icon: ✪.

PROBLEM 1

Page 1: The Story

Keweenaw Rocks

Rebecca grew up in Santa Rosa, California, but one summer she visited her cousin Jason, who lived on the Keweenaw Peninsula in the Upper Peninsula of Michigan (see Figure 6.1). She was a little homesick. She was more than 2,000 miles from home, and much of the landscape was very different from what she was used to. But some of the rocks looked similar to the basalts around Santa Rosa. Where did all the basalt come from? This dark volcanic rock looked similar to the rocks on the West Coast, but on the West Coast they were associated with volcanic activity. But there weren't any volcanoes in Michigan, were there?

There were also sedimentary rocks. Some were sandstones (Jacobsville Sandstone). Others were identified as conglomerates. They contained fist-size pebbles. And the really interesting thing was that these conglomerates were mixed in with the basalts. So there was a layer of volcanic rock, then a layer of conglomerates, then more volcanic rock—and this continued over and over again. How do you explain that? What's the story? What do the rocks tell us about what was going on in Michigan millions of years ago?

✪ **Figure 6.1. Keweenaw Peninsula of Michigan**

Your Challenge: *Help Rebecca understand what she is seeing. Develop a model that reconstructs the geologic history of the Keweenaw Peninsula and explains why there are conglomerates and other sedimentary rocks interlayered with the volcanic rocks in Michigan's Upper Peninsula where there are no volcano cones.*

PROBLEM 1

Page 2: More Information

Keeweenaw Rocks

During her investigation of the geology of the Keweenaw Peninsula, Rebecca looked up several types of information and went to a local museum to learn more about the natural history of the area. She used maps, a geologic timescale, and some field guides to help her identify rocks she found as she explored the peninsula.

One of the people she met was an older man who had worked in the mines and was interested in rocks as a hobby. He showed her several "hand samples," rocks she could hold and take a closer look at. He also told her that the layers of basalts and sedimentary rocks that were visible in the Keweenaw Peninsula continued under Lake Superior and resurfaced at Isle Royale and the western shore of Lake Superior. Geologists could "see" these buried layers, because the iron in the basalt affected Earth's magnetic field.

Take a look at all the information Rebecca gathered while exploring the Keweenaw Peninsula. Photos of the hand sample rocks are on Page 3: Resources.

Your Challenge: *Help Rebecca understand what she is seeing. Develop a model that reconstructs the geologic history of the Keweenaw Peninsula and explains why there are conglomerates and other sedimentary rocks interlayered with the volcanic rocks in Michigan's Upper Peninsula where there are no volcano cones.*

PROBLEM 1

Page 3: Resources

Keweenaw Rocks

Information on Michigan Geology

- Bornhorst, T. J., and W. I. Rose. 1994. "Self-guided geological field trip to the Keweenaw Peninsula, Michigan." Institute on Lake Superior Geology proceedings, 40th annual meeting, Houghton, MI, Vol. 40, Part 2. *www.d.umn. edu/prc/lakesuperiorgeology/Volumes/ILSG_40_1994_pt2_Houghton.cv.pdf.*

- Gillespie, R., W. B. Harrison III, and G. M. Grammer. 2008. *Geology of Michigan and the Great Lakes.* Cengage Brooks/Cole. Also available at *http://custom.cengage. com/regional_geology.bak/data/Geo_Michigan_Watermarked.pdf.*

- Michigan Rocks. 2008. Michigan geologic time line list. *www.educ.msu.edu/ michiganrocks/PDFs/migeol1.pdf.*

- Oxley, P. (Series producer and director). 2015. *Making North America: Origins* [video]. NOVA. Boston: WGBH Educational Foundation.

Maps

- Michigan bedrock map: *http://custom.cengage.com/regional_geology.bak/data/Geo_ Michigan_Watermarked.pdf.*

- This Dynamic Planet (interactive map website): *http://nhb-arcims.si.edu/ ThisDynamicPlanet/index.html.*

- Geologic map of the Keweenaw Peninsula and adjacent area, Michigan (IMAP 2696): *https://pubs.er.usgs.gov/publication/i2696.*

General Geology References

- Press, F., R. Siever, J. Grotzinger, and T. Jordan. 2003. *Understanding Earth.* 4th ed. New York: W. H. Freeman.

- Marshak, S. 2005. *Earth: Portrait of a planet.* 2nd ed. New York: W. W. Norton.

Possible Materials for Modeling Tectonic Plates

- Graham crackers

- Soft frosting, whipped cream, or nondairy whipping cream

- Sprinkles or bread crumbs

Hand Sample Rocks

- Conglomerates (see Figure 6.2)
- Basalts (see Figure 6.3)
- Sandstones (see Figure 6.4)

✪ **Figure 6.2. Conglomerates From Copper Harbor Area**

 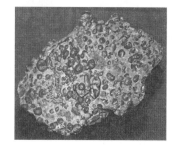

✪ **Figure 6.3. Portage Lake Basalts Outcrop (top) and Hand Sample (bottom)**

✪ **Figure 6.4. Jacobsville Sandstone Hand Samples (top) and Bluffs (bottom)**

Note: Full-color versions of the figures above are available on the book's Extras page at *www.nsta.org/pbl-earth-space*.

PROBLEM 1

Teacher Guide

Keweenaw Rocks

Problem Context

The long geologic history of the Keweenaw Peninsula can be understood by studying the landscape, rock samples, and geologic maps. From careful examination of the rocks, (both texture and composition) that make up the landscape, one can conclude that they represent a divergent boundary, a place where plates move apart (sometimes referred to as a rift zone).

Related Context

The East African Rift valley is a divergent boundary and represents a modern-day rift environment. Comparisons can be made between it and the Keweenaw Rift. Iceland, which is located on the Mid-Atlantic Ridge, is another divergent boundary with active eruptions of basaltic lava through fissures. Iceland is being pulled apart much like the Keweenaw Peninsula was over a billion years ago.

Problem Solution

Model response: Early on in Michigan's history (2,500–1,500 million years ago) tectonic plates came together, forming mountains. The uplifted land was subject to weathering and erosion, and broken bits of it ended up in conglomerates. The conglomerates formed when large broken bits of rock collected close to their source. If the bits had traveled very far, they would have been broken up into smaller pieces and rounded off.

Around 1,200 million years ago, rifts occurred in the area. Rifts usually occur at divergent plate boundaries, but at that time a rift formed in the middle of the North American Plate. Floods of basaltic lava periodically erupted through fissures in a hot spot or thin part of the crust. This type of eruption does not form the tall cone that most people think of when they picture a volcano. Instead, this is considered a shield volcano. Shield volcanoes cover a large area and have a low profile. They form from the fluid lava that flows out rather than building up. The result was basalt interlayered in the conglomerates.

The weight of the basalt caused the thin crust to sag. Between eruptions, sand from the erosion of granites and other igneous rocks was carried into the Keweenaw Rift valley to eventually form (Jacobsville) sandstone.

Activity Guide

While exploring the materials described on the Resources page, there are several ways to engage students in activities that will support their research on this problem. If possible, students should be allowed to hold and examine samples of basalts, conglomerates, and sandstones. Even if the samples differ somewhat from those from the Keweenaw Peninsula, seeing and feeling the differences in the rock types is helpful. Important clues to rocks' identity and history are color (especially the basalts' characteristic dark gray) and grain or crystal size. (Sand grains will be visible in the sandstone. Particles may be visible in rock fragments or clasts in the conglomerate, depending on the rock type. The basalt will have no apparent crystalline structure, having cooled quickly on Earth's surface.)

To build the model, students can use graham crackers on a generous layer of soft frosting, whipped cream, or nondairy whipping cream to simulate the tectonic plate on the mantle. Bread crumbs or candy sprinkles on the graham crackers can represent sedimentary rocks on top of the plates. If students press down gently on either side of two adjoining graham crackers and then gently separate the crackers, the frosting will ooze up onto the crackers. Students can add another layer of sedimentary rocks/crumbs or sprinkles and repeat the process. This works best if the frosting or cream underlayment is very soft and thick. Converging continental plates can be simulated with graham crackers moistened with water and placed on the soft frosting. When two are pushed together, they deform and crumple and pieces pile up, forming a thick region in the crust. What this simulation does not show is that because the crust is floating on the mantle, there is actually more crust below the mountain range than above.

If you choose to include this or a similar simulation activity, an important step not always mentioned in the lesson plans is to compare the appearance of the graham cracker models with photographs and topographical maps (SEP 2: Developing and Using Models), especially for locations like the Keweenaw Peninsula of Michigan, Lassen Volcanic National Park (see Problem 2 in this chapter), or the San Andreas Fault (see Problem 3 in this chapter).

We have provided several high-resolution images in the chapter, including color images in the online resources (*www.nsta.org/pbl-earth-space*). Students can use the images to supplement observation of any hand samples you have in class.

SAFETY PRECAUTIONS

1. Remind students that they should not eat materials provided in the science laboratory.

2. Use caution when handling rock samples. Some may have sharp edges, which can cut or puncture skin.

3. Wash hands with soap and water when the activity is complete.

PROBLEM 1

Assessment

Keweenaw Rocks

Transfer Task

The East African Rift valley of eastern Africa is an example of a divergent plate boundary. Based on what you know about the geology of the Keweenaw Peninsula, what types of rocks would you expect to see in the East African Rift valley? [CC 1: Patterns; CC 2: Cause and Effect: Mechanism and Explanation] Include general rock types you might see, as well as an explanation of the way those rocks formed.

Model response: The East African Rift valley is likely to have dense igneous rocks like basalt and gabbro. These igneous rocks would form as magma floats up from the mantle into the space left as the divergent plates move apart from each other. The igneous rocks would form in thick layers, with newer intrusions of igneous rock closer to the center of the boundary.

I would also expect to see sedimentary rocks on top of the igneous rocks, especially near the center of the rift valley where water and wind can deposit layers of sand, gravel, and silt from the erosion and weathering of higher rocks. Rocks like sandstone, shale, and conglomerate would be most likely. [SEP 2: Developing and Using Models; SEP 6: Constructing Explanations and Designing Solutions]

Problem 2: Lassen's Lessons

Alignment With the *NGSS*

PERFORMANCE EXPECTATIONS	• *MS-ESS2-2:* Construct an explanation based on evidence for how geoscience processes have changed Earth's surface at varying time and spatial scales. • *MS-ESS2-3:* Analyze and interpret data on the distribution of fossils and rocks, continental shapes, and seafloor structures to provide evidence of the past plate motions. • *HS-ESS2-1:* Develop a model to illustrate how Earth's internal and surface processes operate at different spatial and temporal scales to form continental and ocean-floor features.
SCIENCE AND ENGINEERING PRACTICES	• Developing and Using Models • Analyzing and Interpreting Data • Constructing Explanations and Designing Solutions • Obtaining, Evaluating, and Communicating Information
DISCIPLINARY CORE IDEAS	• *ESS2.A: Earth Materials and Systems* 　○ *Grades 6–8:* All Earth processes are the result of energy flowing and matter cycling within and among the planet's systems. This energy is derived from the Sun and Earth's hot interior. The energy that flows and matter that cycles produce chemical and physical changes in Earth's materials and living organisms. • *ESS2.B: Plate Tectonics and Large-Scale System Interactions* 　○ *Grades 6–8:* Maps of ancient land and water patterns, based on investigations of rocks and fossils, make clear how Earth's plates have moved great distances, collided, and spread apart. 　○ *Grades 9–12:* Plate tectonics is the unifying theory that explains the past and current movements of the rocks at Earth's surface and provides a framework for understanding its geologic history. Plate movements are responsible for most continental and ocean-floor features and for the distribution of most rocks and minerals within Earth's crust.
CROSSCUTTING CONCEPTS	• Patterns • Cause and Effect: Mechanism and Explanation • Systems and System Models • Energy and Matter: Flows, Cycles, and Conservation

Keywords and Concepts

Rocks as evidence of past events, plate tectonics, convergent boundary, subduction zone, metamorphic rocks

Problem Overview

Lassen Volcanic National Park has several types of rocks in one location. It is an unusual mix of all three (igneous, sedimentary, and metamorphic). What does each rock type tell us about the area's geologic history?

Images available in full color on the Extras page (*www.nsta.org/pbl-earth-space*) are marked with the following icon: ✪.

PROBLEM 2

Page 1: The Story

Lassen's Lessons

Rebecca returned to her home area of Northern California after a trip to Michigan. In Michigan, she had seen some interesting igneous rocks and conglomerates with streaks of copper running through them. As she drove home, she passed Mount Shasta, a large volcano (see left side of Figure 6.5). She thought it would be fun to explore the volcano, but more fun if she could do it with someone. She invited her cousin Jason to come out and look at rocks in Northern and Central California. Jason lived in the Upper Peninsula of Michigan and, like Rebecca, was a rock hound.

✪ **Figure 6.5. Mount Shasta (left) and Lassen Peak (right)**

After Jason arrived in San Francisco, he and Rebecca took a long drive north along the coast to Crescent City and then inland to Lassen Volcanic National Park to see a recently active volcano. As they drove through Lassen, Jason noted the general lack of black lavas (the basalts often associated with some kinds of volcanoes) and the presence of wide areas with light-color rocks that seemed to be more granular and ashy. They looked like they had "flowed" out of the volcanic peak. Rebecca and Jason stopped along the road, where a ranger told them that this flow formed in 1915 during the last series of eruptions (see Figures 6.6–6.9). Jason also noticed that Lassen, like Shasta, is a distinct volcano with fairly steep sides that prominently stick up above the surrounding area (see Figure 6.5): "Hmm, this looks kind of similar to those pictures of Mount Fuji in Japan in our Earth Science textbook. A lot of volcanoes have the same shape, but the rocks around them are different." What's going on? What's the story?

Your Challenge: *Help Rebecca and Jason explain what they are seeing. Construct a model of the geologic history of Northern California that explains (1) why the volcanic rocks are so diverse and (2) what the relationship between the volcanic and metasedimentary rocks is.*

Figure 6.6. Lassen Peak Before 1914 Eruption

Figure 6.7. 1914 Eruption of Lassen Peak

Figure 6.8. 1915 Eruption of Lassen Peak

✪ Figure 6.9. Bumpass Hell Region, Modern Area of Volcanic Flow

PROBLEM 2

Page 2: More Information

Lassen's Lessons

After leaving the park, Jason and Rebecca took a route home that took them near the coast again. As they drove along the coast, Jason noted the almost complete lack of wide beaches that he had always associated with ocean coastlines. They collected some of the smaller fragments of rock, including chert, basalt, limestone, sandstone, blueschist, and serpentinite. The landforms along the coast consisted of large outcrops and boulders of this very diverse material mixed in with what appeared to be very fine-grained material.

Rebecca had also been thinking. "Jason, when we were at the Keweenaw Peninsula, we found out that the area was a rift—a divergent plate boundary—and that there had been massive flows of relatively 'fluid' basalt. This doesn't look at all similar. Why is the volcanism so different? What's the story here?"

See the rock samples listed on Page 3: Resources for examples of some of these rocks.

Your Challenge: *Help Rebecca and Jason explain what they are seeing. Construct a model of the geologic history of Northern California that explains (1) why the volcanic rocks are so diverse and (2) what the relationship between the volcanic and metasedimentary rocks is.*

PROBLEM 2

Page 3: Resources

Lassen's Lessons

Maps and Geologic Timescale

- Highway maps of the area around Lassen Peak are available at *www.aapg.org/ publications/special-publications/maps/details/articleid/4392/pacific-southwest-region-geological-highway-map.*

- Description of the Phanerozoic eon by M. A. Kazlev is available at *http://palaeos. com/phanerozoic/phanerozoic.htm.*

Books and Educational Materials

- Elder, W. P. 2001. Geology of the Golden Gate headlands. In *Geology and natural history of the San Francisco Bay Area: A field-trip guidebook*, eds. P. W. Stoffer and L. C. Gordon, 61–86. U.S. Geological Survey Bulletin 2188. Reston, VA: U.S. Geological Survey. *http://pubs.usgs.gov/bul/b2188/b2188ch3.pdf.*

- Fichter, L. S. 2000. An introduction to igneous rocks. *http://csmres.jmu.edu/ geollab/Fichter/IgnRx/Introigrx.html#simpleclass.* Harrisonburg, VA: James Madison University.

- Harris, S. L. 2005. Lassen Peak: California's most recently active volcano. In *Fire mountains of the West: The Cascade and Mono Lake volcanoes.* Missoula, MT: Mountain Press.

- Kane, P. S. 1980. *Through Vulcan's eye: The geology and geomorphology of Lassen Volcanic National Park.* Walter Lithograph.

Possible Materials for Modeling Tectonic Plates

- Graham crackers
- Soft frosting, whipped cream, or nondairy whipping cream
- Sprinkles or bread crumbs

Hand Sample Rocks

- Granite (see Figure 6.10, p. 124)
- Basalt (see Figures 6.10, p. 124, and 6.13, p. 125)
- Blueschist (see Figure 6.11, p. 124)

- Andesite (see Figure 6.12)

- Sandstone (see Figure 6.14)

- Chert (see Figure 6.15)

- Serpentinite (see Figure 6.16)

✪ Figure 6.10. Volcanic Boulders of Granite (left) and Basalt (right) at Lassen Volcanic National Park

✪ Figure 6.11. Blueschists Near California Coast

✪ Figure 6.12. Andesite Samples

Note: Full-color versions of the figures above are available on the book's Extras page at *www.nsta.org/pbl-earth-space*.

⚙ **Figure 6.13. Pillow Basalt Formation in Marin County, California**

⚙ **Figure 6.14. Greywacke Sandstone**

⚙ **Figure 6.15. Ribbon Chert, With Folded Layers Along the California Coast**

⚙ **Figure 6.16. Serpentinite Sample**

Note: Full-color versions of the figures above are available on the book's Extras page at *www.nsta.org/pbl-earth-space*.

PROBLEM 2

Teacher Guide

Lassen's Lessons

Problem Context

The Lassen Volcanic National Park area of Northern California represents an environment with an interesting geologic history. Mount Shasta indicates a volcanic past, and in the park there are many examples of igneous rocks (basalts and andesites). Volcanoes thrust rocks up where they are subject to weathering and erosion. This gives rise to sediments that may be altered by the heat and pressure (metasedimentary rocks like schist and serpentinite that are metamorphosed sedimentary rocks) and sedimentary rocks that were formed in an oceanic environment (chert and greywacke).

Lassen can be compared to other places in the world where subduction is taking place, such as Japan or the west coast of South America.

Problem Solution

Model response: The geology of Northern California reflects a history of violent volcanic and tectonic activity. The presence of an active volcano is evidence of this. Volcanoes like Lassen Peak and Mount Shasta form near subduction zones. Two plates collide there, with the less dense, continental North American Plate riding up and over the denser, oceanic Pacific Plate. As the oceanic plate sinks under, the continental plate is uplifted, folded, and crumbled. Water from the subducting slab combined with the rocks above it gives rise to a mix that melts at lower temperatures. This allows heat from the mantle to melt the slab, creating lava that rises toward the surface, forming volcanoes. The lava from these volcanoes may contain different substances than either of the original plates due to the complex chemistry going on as the subducting plate brings water with it into the mantle.

In subduction zones, a wide variety of sedimentary rocks may form from eroded material from the volcanoes, the uplift and exposure of the continental margin to erosion and weathering, and accretion (scraping off) from the oceanic plate.

The heat and pressure also cause the sedimentary rocks (such as sandstone) and igneous rocks (such as andesite and basalt) at these boundaries to recrystallize and deform. The results are metamorphic rocks, like the metamorphosed chert, serpentinite, and blueschist. These rocks are often found in folded layers created by the pressure of the subduction zone, like those seen in the photographs of the ribbon chert and blueschist. The rocks found along the coast provide evidence of this plate movement.

Activity Guide

While exploring the materials described in the Resources page, there are several ways to engage students in activities that will support their research on this problem. If possible, students should be allowed to hold and examine samples of the rocks listed on Page 3. Even if the samples differ somewhat from those from California, seeing and feeling the differences in the rock types is helpful. Important clues to rocks' identity and history are color (especially the basalts' characteristic dark gray versus the lighter-colored andesite) and crystal size. (Sand grains will be visible in the sandstone. The basalt will have no apparent crystalline structure, having cooled quickly on Earth's surface. The grains in schists will be flattened and aligned by the pressure of metamorphism.)

To build the model, students can use graham crackers on a generous layer of soft frosting, whipped cream, or nondairy whipping cream to simulate the tectonic plate on the mantle. Dip half of one graham cracker in water to represent the continental plate. Bread crumbs or candy sprinkles on the graham crackers can represent sedimentary rocks on top of the plates. Students can simulate subduction by pushing one cracker (oceanic plate) under the other, making sure to scrape the sprinkles or crumbs off the top of the subducting plate with the upper cracker. The pileup of the sprinkles/crumbs, along with crumbs breaking off the upper crust, represents the mix of sedimentary rocks associated with subduction zones. It is difficult with a physical model to show the metamorphism and volcanic events associated with subduction zones. However, the characteristic trenches are visible at the edge of the upper cracker.

If you choose to include this or a similar simulation activity, an important step not always mentioned in the lesson plans is to compare the appearance of the graham cracker models with photographs and topographical maps (SEP 2: Developing and Using Models), especially for locations like Lassen Volcanic National Park, the San Andreas Fault (see Problem 3 in this chapter), or the Keweenaw Peninsula of Michigan (see Problem 1 in this chapter).

We have provided several high-resolution images in the chapter, including color images in the online resources (*www.nsta.org/pbl-earth-space*). Students can use the images to supplement observation of any hand samples you have in class.

SAFETY PRECAUTIONS

1. Remind students that they should not eat materials provided in the science laboratory.

2. Use caution when handling rock samples. Some may have sharp edges, which can cut or puncture skin.

3. Wash hands with soap and water when the activity is complete.

PROBLEM 2

Assessment

Lassen's Lessons

Transfer Task

Rebecca's favorite vacation was a trip to Puerto Rico. One of the highlights was the drive through the mountains near San Germán. The tour van had a flat tire on one of the highways. While the driver changed the tire, Rebecca noticed that the hillside along the highway had three distinct layers of rock. Near the road, there was a dark layer of basalt, and just above it were folded layers of schist. The top layer was ancient limestone. This reminded her of another area she had seen in California.

Based on the rocks she found so close together, what can you tell about the geologic history of the mountains of western Puerto Rico? Explain your answer.

> *Model response:* The combination of schist and basalt (metamorphic and igneous) rocks suggests a subduction zone. In fact, Puerto Rico sits on a location where three tectonic plates converge, as evidenced by the three distinct rock types. There is a subduction zone along one of the plate boundaries. The top layer of limestone was likely created in the oceans, and when the denser plate collided with it, the limestone was pushed upward. As the denser plate sank deeper, magma from the mantle floated up into to the subduction zone to create the basalt at the roadside. The heat and pressure of the plates rubbing against each other caused the layers to fold and partially melt, forming the folded schist in the middle layer. [SEP 2: Developing and Using Models; CC 1: Patterns; CC 2: Cause and Effect: Mechanism and Explanation]

Problem 3: San Andreas

Alignment With the *NGSS*

PERFORMANCE EXPECTATIONS	• *MS-ESS2-2:* Construct an explanation based on evidence for how geoscience processes have changed Earth's surface at varying time and spatial scales.
	• *MS-ESS2-3:* Analyze and interpret data on the distribution of fossils and rocks, continental shapes, and seafloor structures to provide evidence of the past plate motions.
	• *HS-ESS1-5:* Evaluate evidence of the past and current movements of continental and oceanic crust and the theory of plate tectonics to explain the ages of crustal rocks.
SCIENCE AND ENGINEERING PRACTICES	• Developing and Using Models
	• Analyzing and Interpreting Data
	• Constructing Explanations and Designing Solutions
	• Obtaining, Evaluating, and Communicating Information
DISCIPLINARY CORE IDEAS	• *ESS2.A: Earth Materials and Systems*
	○ *Grades 6–8:* All Earth processes are the result of energy flowing and matter cycling within and among the planet's systems. This energy is derived from the Sun and Earth's hot interior. The energy that flows and matter that cycles produce chemical and physical changes in Earth's materials and living organisms.
	• *ESS2.B: Plate Tectonics and Large-Scale System Interactions*
	○ *Grades 6–8:* Maps of ancient land and water patterns, based on investigations of rocks and fossils, make clear how Earth's plates have moved great distances, collided, and spread apart.
	○ *Grades 9–12:* Plate tectonics is the unifying theory that explains the past and current movements of the rocks at Earth's surface and provides a framework for understanding its geologic history. Plate movements are responsible for most continental and ocean-floor features and for the distribution of most rocks and minerals within Earth's crust.
CROSSCUTTING CONCEPTS	• Patterns
	• Cause and Effect: Mechanism and Explanation
	• Systems and System Models
	• Energy and Matter: Flows, Cycles, and Conservation

Keywords and Concepts

Plate tectonics, transform boundary, subduction zone, strike-slip fault

Problem Overview

Southern California is affected by the San Andreas Fault. The geologic features and rock types along this fault tell of a history very different from the region just a short distance to the north. Two cousins try to learn more about how that part of California differs from the northern part of the state.

Images available in full color on the Extras page (*www.nsta.org/pbl-earth-space*) are marked with the following icon: ✪.

PROBLEM 3

Page 1: The Story

San Andreas

Rebecca lived in California, and her cousin Jason was visiting from Michigan. They were driving south back to San Francisco along Interstate 5 after visiting the volcanic Mount Shasta. Jason noted that there were no more volcanoes once they got south of Lassen Volcanic National Park. Cutting back to the coast, they stopped off at Point Reyes National Seashore. When they stopped at the visitor center, the ranger suggested they hike the Earthquake Trail that started across the parking lot.

Jason saw posts that marked the line of the San Andreas Fault (see Figure 6.17) and had his picture taken straddling the fault line. As they walked the trail, Rebecca and Jason noted that the park service had put out samples of rocks found on either side of the fault—with rocks similar to those they saw farther north on the east side of the fault, including fine-grained sedimentary rocks like shale, sandstone, and chert. On the west side, they saw mostly granites. Jason thought this rather odd. Why should the rocks just a few feet apart be so different? What is the story here?

✪ Figure 6.17. The San Andreas Fault

Your Challenge: *Help Rebecca and Jason understand what they are seeing. Construct a model that explains (1) why there are no active volcanoes south of Lassen Volcanic National Park and (2) why there are jumbled sedimentary rocks on the east side of the fault, but not on the west side.*

PROBLEM 3

Page 2: More Information

San Andreas

As Jason and Rebecca walked the trail, they crossed the fault and moved to the west side. The rocks looked very different from those on the east side of the fault—they were granites with distinct crystals of different colors compared with the chert and pillow basalts on the east side, which had no apparent crystals. Along the boundary, they found fine-grained schist with layers of igneous granite embedded in it.

As they walked the trails of the park, they found a spot where a fence had been broken, and ends that used to meet were now a few yards apart (see Figure 6.18).

✪ **Figure 6.18. Earthquake Trail Fence**

When they returned home, they looked back at their photographs from the trip and compared them to information from a National Park Service map of the area where they walked in Point Reyes and a book they found in the library called *A Land in Motion* (by Michael Collier) to see if they could figure out how igneous rocks and metamorphic rocks and sedimentary rocks could be found so close together, but with such a clear boundary between them, as seen in the road cut they passed on the highway (see Figure 6.19).

Your Challenge: *Help Rebecca and Jason understand what they are seeing. Construct a model that explains (1) why there are no active volcanoes south of Lassen Volcanic National Park and (2) why there are jumbled sedimentary rocks on the east side of the fault, but not on the west side.*

✪ **Figure 6.19. The San Andreas Fault Visible at Road Cut**

References

Collier, M. 1999. *A land in motion: California's San Andreas Fault.* Oakland, CA: University of California Press.

National Park Service. n.d. Maps [of Point Reyes National Seashore]. *www.nps.gov/pore/planyourvisit/maps.htm.*

Note: Full-color versions of Figures 6.18 and 6.19 are available on the book's Extras page at *www.nsta.org/pbl-earth-space.*

PROBLEM 3

Page 3: Resources

San Andreas

Books

Collier, M. 1999. *A land in motion: California's San Andreas Fault.* Oakland, CA: University of California Press.

Press, F., R. Siever, J. Grotzinger, and T. Jordan. 2003. *Understanding Earth.* 4th ed. New York: W. H. Freeman.

Marshak, S. 2005. *Earth: Portrait of a planet.* 2nd ed. New York: W. W. Norton.

Websites

- Indiana University, Indiana Geological Survey. n.d. Study of faults and earthquakes with foldable fault blocks. *https://igs.indiana.edu/lessonplans/faultblock.pdf.*

- U.S. Geological Survey. 2014. Understanding plate motions. *http://pubs.usgs.gov/gip/dynamic/understanding.html.*

Possible Materials for Modeling Tectonic Plates

- Graham crackers
- Soft frosting, whipped cream, or nondairy whipping cream
- Sprinkles or bread crumbs

Hand Sample Rocks

- Granite (see Figure 6.20, p. 134)
- Chert (see Figure 6.21, p. 134)
- Pillow basalt (see Figure 6.22, p. 134)
- Shale (see Figure 6.23, p. 134)
- Sandstone (see Figure 6.24, p. 134)

✪ Figure 6.20. Granite

✪ Figure 6.21. Sedimentary Chert Outcrop (top) and Sample (bottom)

✪ Figure 6.22. Pillow Basalt

✪ Figure 6.24. Sandstone

✪ Figure 6.23. Shale

Note: Full-color versions of the figures above are available on the book's Extras page at *www.nsta.org/pbl-earth-space*.

PROBLEM 3

Teacher Guide

San Andreas

Problem Context

A walk along the San Andreas Fault shows some unusual rock samples, with igneous, sedimentary, and metamorphic rocks showing clear patterns in their location that may help explain what kind of fault is found here. Comparing the San Andreas Fault with the Keweenaw Peninsula and Lassen Volcanic National Park will reveal some important differences in the types of geologic events and the evidence that illustrates how plate boundaries can change over time.

Problem Solution

Model response: The combination of basalt, sedimentary rocks, and schist indicates a subduction zone. The granites to the west appear to be unrelated to this subduction zone. Their presence and the displaced fence indicate a transform fault, a place where plates slide past each other rather than colliding into each other. This type of fault is sometimes called a strike-slip fault and does not produce volcanoes, because there is no subduction and melting of plates—hence the lack of volcanoes south of Lassen.

The two plates meet along the San Andreas Fault. The Pacific Plate is to the west and is made of granites, igneous rocks that formed slowly, as evidenced by the large crystals. The North American Plate is to the east, and it is made of less dense sedimentary rocks like chert, shale, and sandstone.

Along the boundary of these two plates, the Pacific Plate moves northward, relative to the North American Plate. The friction between the plates resists that movement, and they "stick" to each other. When the plates slip, the result is an earthquake. This is why the region from San Francisco to Los Angeles has frequent tremors. [CC 2: Cause and Effect: Mechanism and Explanation]

[*Note to teacher:* The main ideas are contained in the first three paragraphs, but you may choose to have your students go further by drawing more detailed parallels between this case and the case of Lassen Peak in Problem 2. This evidence can then be used to examine how plate boundaries change over time (SEP 6: Constructing Explanations and Designing Solutions).]

Closer examination of the rocks in the Keweenaw Peninsula reveals a tectonic history similar to the plate movement suggested by the evidence at the present-day Lassen Volcanic National Park. [SEP 4: Analyzing and Interpreting Data] The area was once a convergent boundary. The less dense North American Plate slid over the top of the denser Pacific Plate, creating a subduction zone. This created volcanoes like Lassen Peak and Mount Shasta to the north, but it also created areas where the rocks were placed under very high pressure and heat for a long period of time. This is what created the schists Rebecca and Jason saw. The embedded granites are the pieces of the Pacific Plate that were pushed into the melting sedimentary rocks. This mix of metamorphic and sedimentary rock gives these rocks the label *metasedimentary.*

Activity Guide

While exploring the materials described in the Resources page, there are several ways to engage students in activities that will support their research on this problem. If possible, students should be allowed to hold and examine samples of the rocks listed on Page 3. Even if the samples differ somewhat from those from California, seeing and feeling the differences in the rock types is helpful. Important clues to rocks' identity and history are color (especially the basalts' characteristic dark gray versus the lighter, mixed colors of granite) and crystal size. (Sand grains will be visible in the sandstone. The basalt will have no apparent crystalline structure, having cooled quickly on Earth's surface. The crystals in schists will be flattened and aligned by the pressure of metamorphism.)

We have provided several high-resolution images in the chapter, including color images in the online resources (*www.nsta.org/pbl-earth-space*). Students can use the images to supplement observation of any hand samples you have in class.

To build the model, students can use graham crackers on a generous layer of soft frosting, whipped cream, or nondairy whipping cream to simulate the tectonic plate on the mantle. Bread crumbs or candy sprinkles on the graham crackers can represent sedimentary rocks on top of the plates. Students can simulate the transform fault by scraping the edges of two adjacent plates together, moving one plate away from the body and the other plate toward the body. They will feel the two plates catch on each other and then slip, simulating an earthquake. Most of the sedimentary rocks (crumbs/sprinkles) will remain on their respective plates.

If you choose to include this or a similar simulation activity, an important step not always mentioned in the lesson plans is to compare the appearance of the graham cracker models with photographs and topographical maps (SEP 2: Developing and Using Models), especially for locations like the San Andreas Fault, Lassen Volcanic National Park (see

Problem 2 in this chapter), or the Keweenaw Peninsula of Michigan (see Problem 1 in this chapter).

Students can also build models of strike-slip faults and other types of plate boundaries using the directions in activities such as the Indiana Geological Survey lesson plan on faults and earthquakes listed on Page 3: Resources.

SAFETY PRECAUTIONS

1. Remind students that they should not eat materials provided in the science laboratory.

2. Use caution when handling rock samples. Some may have sharp edges, which can cut or puncture skin.

3. Wash hands with soap and water when the activity is complete.

PROBLEM 3

Assessment

San Andreas

Transfer Task

Jason's father is a geologist. He has a saying about earthquake activity: "For the Americas, east coasts are safer than west coasts." Use maps of the tectonic plates and earthquake activity to check out this saying. Is it true for North, Central, and South America? Explain why or why not.

Model response: It is true that the east coasts of North and South America have very few earthquakes. This is because these coasts are not at plate boundaries. They are in the middle of plates where there is little tectonic activity. On the other hand, the west coasts of these two continents are at convergent boundaries where there is a lot of seismic and tectonic activity. Central America is different. The narrow strip of land is on a small plate between two convergent boundaries. Much of southern Central America is subject to earthquakes.

Rock Cycle and Plate Tectonics Problems: General Assessment

The general and application questions in this section are written to address concepts spanning all three problems in this chapter. You can use these questions in the general assessment whether you choose to use all three problems in a class or use only one or two problems that focus on selected concepts.

General Question

It is said that the history of Earth is written in the rocks. How do rocks tell the geologic history of an area? Include the rock type, age, texture of the rock, the general composition, relationship to other rocks in the environment, and the rock-forming processes involved.

> *Model response:* The history of the formation of Earth's crust can be determined by observing several kinds of evidence. This is possible if we assume that geologic processes work the same way now as they did in the past. Stratigraphy (examining layers of rock) shows the sequence in which different kinds of rocks formed. The fossils found in layers can also give information about the age of the rock, as well as the environment in which the layer was formed. For instance, fossils of mollusks and coral mean that the layer formed in a warm, shallow ocean, while fossilized bones of wooly mammoths suggest that the area was near glaciers during the Ice Age. [SEP 4: Analyzing and Interpreting Data; SEP 6: Constructing Explanations and Designing Solutions]
>
> The types of igneous, metamorphic, and sedimentary rocks also tell us about geologic and environmental conditions at the time the rocks were formed. If crystals in granite are large, the magma cooled slowly, probably below the surface of Earth. If the crystals are extremely small and the rock looks like glass, the lava cooled quickly, possibly after emerging on Earth's surface or dropping into water. The sediments that form a sedimentary rock also give information. For example, shales form from fine, silty sediments like mud in a riverbed, but sandstone still shows grains of the sands from a beach or a dune. If schists and gneiss are found, we know the area once experienced high temperature and pressure that modified the preexisting rock. This could be deep down in the region of the lower crust or at a convergent boundary or subduction zone. The structure and composition of rocks tells us a lot about how, where, and when they were formed. [SEP 4: Analyzing and Interpreting Data]

CHAPTER 6

Application Question

The theory of plate tectonics proposes that plates come together at convergent plate boundaries and spread apart at divergent boundaries. For each of the following boundaries, explain how all the parts of the rock cycle happen:

a. The East African Rift valley (a divergent boundary)

b. Japan (a convergent boundary)

Model response:

a. In the East African Rift valley, igneous rocks are being formed as magma floats into the space left as two tectonic plates separate. The magma cools to form bedrock. In the valley usually left between the two plates, water and wind erode rocks, creating sediments that collect in the lower areas, and these sediments can eventually be cemented together by pressure and dissolved chemicals to become sedimentary rocks. Layers of these rocks can build on top of the bedrock over time. At the boundary where magma touches or comes close to sedimentary rocks, the rocks can melt and be transformed by the pressure. When they cool again, they become metamorphic rock like the schist found in the Keweenaw Peninsula and along the San Andreas Fault. [CC 1: Patterns; CC 2: Cause and Effect: Mechanism and Explanation]

b. In the convergent boundaries like the ones in Japan, a dense oceanic plate meets and sinks under a less dense continental plate. As it sinks deeper toward the mantle, a subduction zone forms. [CC 1: Patterns] At this boundary, layers of sedimentary rocks form from the erosion of the volcanoes, the uplift of the edge of the continental plate, and the sediments scraped off the subducting oceanic plate. As rocks sink below the surface, the pressure of the converging plates and the heat of friction and magma cause the rocks to melt and change to become metamorphic rocks. The heat and pressure in this area can cause magma to rise up toward the surface. [CC 2: Cause and Effect: Mechanism and Explanation] In oceanic-continental convergent zones, these areas usually occur inland from the boundary and eventually emerge in the continental plate to form volcanoes. The magma that breaks the surface will cool quickly without forming crystals. These rocks will eventually be eroded to form more sedimentary rocks. [CC 1: Patterns] However, magmas that cool below ground in fissures are likely to cool more slowly, forming larger crystals like pumice, scoria, and obsidian.

Common Beliefs

Indicate whether the following statements are true (T) or false (F), and explain why you think so *(model responses shown in italics).*

1. The type and location of plate boundaries change over time. *(T) Plate boundaries can change with time. For example, the San Andreas Fault has developed along a former subduction zone, and the Keweenaw Fault went from extension to compression.*

2. Magma that flows out on Earth as lava originates at Earth's core. *(F) Partial melting occurs in the upper mantle, creating magma at subduction zones and rift zones. Thus, the movement of plates influences this process.*

3. The type and nature of volcanism depend on the type of plate boundary. *(T) Volcanic rocks at subduction zones are different from those at rifts.*

4. Rocks can be clearly classified as igneous, metamorphic, or sedimentary. *(F) The rock cycle represents an idealized model of rock-forming processes. In reality, these processes can occur on a continuum of change. For example, one can find a "soft" slate like shale or a "hard" shale like slate.*

5. The sequence in rock formation as illustrated in the rock cycle is igneous to metamorphic to sedimentary rock. *(F) The rock cycle represents a model of processes (cooling and crystallization, weathering and erosion, sedimentation and lithification, and metamorphism). Plate tectonic processes influence the sequence of events.*

6. The location and ages of rocks can indicate the large-scale tectonic history of a region. *(T) Spatial relationships and ages of rocks can tell us about the large-scale tectonic history of a region. For example, the basalts intermingled with conglomerates in the Keweenaw Peninsula indicate an ancient rift zone.*

7. At the San Andreas Fault, a piece of California will eventually fall into the ocean. *(F) The San Andreas Fault represents a transform boundary. Transform boundaries are those along which plates slide by each other.*

WEATHER

The problems in this chapter focus on concepts related to the seasons, water cycle, and weather and on the ways scientists use data to forecast the weather. The concepts included are the impact of the tilt of Earth's axis on the seasons, how gravity and uneven heating move and change water in the water cycle, and patterns in the movement and interactions of air masses.

The chapter addresses the crosscutting concepts (CCs) described in the *Next Generation Science Standards* (*NGSS*; NGSS Lead States 2013), especially Patterns (CC 1); Cause and Effect: Mechanism and Explanation (CC 2); Systems and System Models (CC 4); and Energy and Matter: Flows, Cycles, and Conservation (CC 5). The problems help students develop the scientific practices of Developing and Using Models (*NGSS* science and engineering practice [SEP] 2), Analyzing and Interpreting Data (SEP 4), Constructing Explanations and Designing Solutions (SEP 6), and Obtaining, Evaluating, and Communicating Information (SEP 8). Disciplinary core ideas (DCIs) emphasized in the chapter include Earth and the Solar System (ESS1.B), The Roles of Water in Earth's Surface Processes (ESS2.C), and Weather and Climate (ESS2.D).

Big Ideas
Seasons

- Earth's seasons are caused by the tilt of Earth's axis. This axis is not perpendicular to Earth's orbital plane; instead, it is tilted 23.5° with respect to the vertical and maintains that orientation in space for long periods of time (millennia).

- The tilt of the axis causes the Northern Hemisphere to experience more intense sunlight and more hours of daylight when that hemisphere is tilted toward the Sun. This increase in intensity and duration warms that part of Earth, which warms the air and results in the Northern Hemisphere's summer. Conversely, the decrease in the intensity and duration of sunlight six months later, when Earth has moved to the opposite side of the Sun and the Northern Hemisphere is tilted away from the Sun, results in that hemisphere's winter.

- In the Southern Hemisphere, the seasons are opposite from those in the Northern Hemisphere.

- Equatorial regions experience less variation in sunlight intensity and duration over the course of a year; in contrast, high-latitude locations, such as the Arctic (and Antarctic), experience more dramatic changes in the angle of the Sun and the duration of sunlight over the course of a year.

Water Cycle

- Water can exist in three different states or phases: solid (ice), liquid, and gas (vapor). The liquid and solid water that is above ground is visible to us as snow, ice, glaciers, surface water, oceans, rain, and clouds. But water also exists in the soil and rocks and as invisible water vapor in the atmosphere.

- Water moves continuously around Earth. The processes that move and change water are driven by the Sun's radiation and the force of gravity.

- Gravity moves liquid water and solid water in glaciers downhill. Gravity also pulls precipitation such as rain and snow from the sky to the surface of Earth.

- Heat is required to move water up against the force of gravity. If water absorbs enough heat, it will evaporate into the atmosphere.

- When liquid water evaporates or a solid melts, heat is absorbed from the surroundings. Freezing and condensation return heat to the surroundings. In this way, water moves energy around Earth.

Air Masses and Weather

- Large air masses are responsible for much of the weather people experience. These masses take on the general temperature and moisture content of the land or water over which they develop.

- The edges of air masses are visible on weather maps as abrupt changes in temperature, humidity, or wind patterns.

- Due to unequal heating of Earth, there are large differences in the characteristics of the air masses. Even though the intensity of the Sun's light is constant, Earth is heated unevenly, because (1) sunlight does not fall at the same angle and for the same duration on all of Earth's surfaces; (2) clouds reflect sunlight, preventing it from reaching Earth's surface; and (3) different areas absorb and reflect different amounts of sunlight.

- General heating at the equator and cooling at the poles cause density differences in the air and ocean water at those latitudes. These density differences lead to large convection currents in both the air and water. Examples of these are the prevailing westerly winds in the United States and the cool current of water that flows from south to north along the East Coast of the United States.

- Weather events occur when air masses with different characteristics meet. The air mass moving into an area has a leading edge or front.

- Generally, when a cold front encounters a warm air mass, storms are generated, followed by clear, cooler weather. When a warm front overtakes a cold front, the weather is generally overcast with light precipitation, followed by warm, humid weather. Severe weather occurs when air masses with extreme differences in temperature and moisture meet.

Conceptual Barriers
Common Problems in Understanding

Many students focus on local weather forecasts without looking at the effects of large air masses, global air and water currents, or particular land formations. In addition, it is difficult to represent the many variables and processes involved in weather:

- Standard weather maps do not convey the differences in atmospheric conditions at different altitudes.

- It is difficult for two-dimensional representations to convey the seasonal changes in the path of the Sun and the dependence of the Sun's path on the latitude of the observer.

- Many representations of the water cycle fail to show rain over the oceans or the connections between groundwater and surface water.

Common Misconceptions

COMMON MISCONCEPTIONS ABOUT SEASONS

- The seasons are caused by variations in Earth's distance from the Sun—summer is when Earth is closer to the Sun, and winter is when Earth is farther away.

- The seasons are caused by changes in the tilt of Earth's axis.

- The hottest day is the first day of summer, and the hottest time of the day is noon.

- The Sun follows the same path across the sky each day.

- The Sun is directly overhead at noon for all observers.

- The hottest temperatures occur in the tropics.

- The tropics have longer days than the temperate zone in summer.

COMMON MISCONCEPTIONS ABOUT THE WATER CYCLE

- The water cycle involves only the freezing and melting of water.

- Once water is used by plants or animals, it's gone.

- Water only evaporates from the ocean or lakes.

- Water in open containers or from open water changes into air or disappears.

- Condensation is when air turns into a liquid.

- Clouds are sponges that hold water.

- Clouds and steam are water vapor.

- Steam is air, heat, or smoke.

- There is no energy transfer involved when water changes phase.

COMMON MISCONCEPTIONS ABOUT AIR MASSES AND WEATHER

- *H* on a weather map means hot; *L* means cold.

- High-pressure systems are dense because they contain a lot of water.

- Meteorologists can fairly accurately predict the weather for two weeks or more.

- Air masses are very small and affect small areas of the country.

Interdisciplinary Connections

There are a number of interdisciplinary connections that can be made with the problems in this chapter. For example, connections can be made to literature, language arts, geography, mathematics, and art and music (see Box 7.1).

Box 7.1. Sample Interdisciplinary Connections for Weather Problems

- **Literature:** Read one of several books about life in Alaska. Examples include *Arctic Lights, Arctic Nights* by Debbie Miller (2003) and *Julie of the Wolves* by Jean Craighead George (1972).

- **Language arts:** Write a story or poem or produce a video story or a photo storyboard of life in Alaska or somewhere the seasons and/or weather are very different from your hometown. Longfellow's *The Song of Hiawatha* (1855/2011) can serve as a model for a poem.

- **Geography:** Study the latitudes of U.S. cities that you are familiar with and compare these with foreign cities in the news.

- **Mathematics:** Use different types of graphs to represent the distribution of water on Earth. Numbers are available from the U.S. Geological Survey at *http://water.usgs.gov/edu/watercycle.html*.

- **Art and music:** Create pictures of the same landscape at different seasons and under different weather conditions. Listen to Antonio Vivaldi's *The Four Seasons* and choose your own music to represent different seasons or different weather.

References

George, J. C. (illustrated by J. Schoenherr). 1972. *Julie of the wolves.* New York: Harper & Row.

Longfellow, H. W. 1855/2011. *The song of Hiawatha: An epic poem.* Oxford, U.K.: Benediction Classics.

Miller, D. S. (illustrated by J. Van Zyle). 2003. *Arctic Lights, Arctic nights.* New York: Walker.

NGSS Lead States. 2013. *Next Generation Science Standards: For states, by states.* Washington, DC: National Academies Press. *www.nextgenscience.org/next-generation-science-standards.*

Problem 1: Northern Lights

Alignment With the *NGSS*

PERFORMANCE EXPECTATIONS	• *1-ESS1-2:* Make observations at different times of the year to relate the amount of daylight to the time of year.
	• *3-ESS2-1:* Represent data in table and graphical displays to describe typical weather conditions expected during a particular season.
	• *5-ESS1-2:* Represent data in graphical displays to reveal patterns of daily changes in length and direction of shadows, day and night, and the seasonal appearance of some stars in the night sky.
SCIENCE AND ENGINEERING PRACTICES	• Developing and Using Models
	• Analyzing and Interpreting Data
	• Constructing Explanations and Designing Solutions
	• Obtaining, Evaluating, and Communicating Information
DISCIPLINARY CORE IDEAS	• *ESS1.B: Earth and the Solar System*
	○ *Grades K–2:* Seasonal patterns of sunrise and sunset can be observed, described, and predicted.
	○ *Grades 3–5:* The orbits of Earth around the Sun and of the Moon around Earth, together with the rotation of Earth about an axis between its North and South Poles, cause observable patterns. These include day and night; daily changes in the length and direction of shadows; and different positions of the Sun, Moon, and stars at different times of the day, month, and year.
	• *ESS2.D: Weather and Climate*
	○ *Grades 3–5:* Climate describes a range of an area's typical weather conditions and the extent to which those conditions vary over years.
CROSSCUTTING CONCEPTS	• Patterns
	• Cause and Effect: Mechanism and Explanation
	• Systems and System Models
	• Energy and Matter: Flows, Cycles, and Conservation

Keywords and Concepts

Seasons, motion of Earth around the Sun

Problem Overview

A boy is worried that his new home in Alaska cannot possibly have warm weather in the summer.

Images available in full color on the Extras page (*www.nsta.org/pbl-earth-space*) are marked with the following icon: ⊙.

PROBLEM 1

Page 1: The Story

Northern Lights

Simon just found out that his family is moving from Iowa to Alaska for a year. They are going to live in Fairbanks, Alaska, which is about 115 miles (185 km) south of the Arctic Circle at 65°N. The Arctic Circle! Does that mean it's going to be cold and snowy all year round? And dark? Isn't it dark all the time that far north? How is Simon ever going to play outside? But maybe he will see the northern lights (see Figure 7.1) more often!

Simon's parents assure him that Fairbanks experiences summer as well as winter. It is home to people, animals, and plants that experience a wide range of temperatures. People enjoy warm summer days, yet they have to bundle up in parkas to step outside of their snug, well-insulated homes in winter. Animals spend their summer days filled with activity, eating and busy with the work of raising their young. Animals' fur coats become light in weight and dark in color. In winter, many animals migrate or spend their time in hibernation. Those that do brave the cold temperature often wear a coat of white. Plants grow quickly and abundantly in the warm summer, often completing their entire life cycle in a few short months. During the cold winter, only the conifers keep their green leaves.

Simon is not convinced that there will be summer weather. If the winter is so cold and dark, how can the summer be so pleasant?

Your Challenge: *Help Simon understand how the temperatures can change so much from summer to winter in Alaska. Create a model that illustrates this phenomenon.*

✪ Figure 7.1. Northern Lights

Note: A full-color version of this figure is available on the book's Extras page at *www.nsta.org/pbl-earth-space*.

PROBLEM 1

Page 2: More Information

Northern Lights

Simon lives in Sioux City, Iowa, which is about 1,660 miles (2,672 km) south of the Arctic Circle at 42°N. Like Fairbanks, Sioux City is home to many people, animals, and plants. People living in Sioux City experience a wide range of temperatures throughout the year, but not the extremes of Fairbanks. Iowa animals and plants also have adaptations to deal with the change in temperature, even though the temperature doesn't vary as much as in Alaska.

Simon has a cousin who lives in Nashville, Tennessee, which is at 36°N and about 830 miles (1,336 km) southeast of Sioux City. Simon knows that the summers and winters are warmer in Nashville than in Sioux City. It looks like the farther north you go, the colder each season is. Doesn't that mean that Fairbanks is going to be very cold?

Simon knows that the seasons are NOT caused by Earth's distance from the Sun. Earth goes around the Sun in an elliptical path that is not quite a circle. It also spins on its own axis. Earth is slightly closer to the Sun in January than in July. In the Northern Hemisphere, we normally experience cold weather in January when we are closer to the Sun and warm weather in July when we are farther away. So what does cause the changing seasons we see on our planet (see Figure 7.2), and why isn't summer in Fairbanks really cold?

Your Challenge: *Help Simon understand how the temperatures can change so much from summer to winter in Alaska. Create a model that illustrates this phenomenon.*

✪ Figure 7.2. Earth From Space

PROBLEM 1

Page 3: Resources and Investigation

Northern Lights

Resources

- Find average monthly temperatures for anywhere in the continental United States at *https://weather.com/maps/averages/normal-temperature.*

- Look up average temperatures and rainfalls for U.S. cities at *http://countrystudies. us/united-states/weather.*

- View a simulation of the Sun's path as seen from anywhere on any date, including sunrise and sunset times, at *www.sunearthtools.com/dp/tools/pos_sun.php.*

Modeling Investigation: Summer and Winter in the Northern Hemisphere
MATERIALS

- For Earth, use a Styrofoam ball on a wooden skewer or pencil (representing Earth's axis) with a nonlatex elastic band perpendicular to the axis (representing the equator).

- Use a pushpin stuck in the foam ball to represent an observer. To represent Simon in Alaska, the pushpin should be at 65° N. You can use a protractor to measure this angle from the "equator" (the nonlatex elastic band).

- For the Sun, use lamps without shades or bright flashlights.

- You can also use the Solar Motion Demonstrator (see Figure 7.3), a handheld device that indicates the path of the Sun for anytime of the year or latitude. See the Extras page at *www.nsta.org/pbl-earth-space* for a pattern and directions.

PROCEDURE

To help Simon understand what he's in for in Alaska, you will need to model for him the situation in both summer and winter. To start, make sure you can model how Earth maintains its tilt as it orbits the Sun. How will you use this model to explain to Simon what to expect in Fairbanks?

SUMMER IN THE NORTHERN HEMISPHERE

✪ Figure 7.3. Solar Motion Demonstrator

1. Hold your foam ball so that it represents Earth at the summer solstice—the longest day of the year in the Northern Hemisphere. Which part of Earth is experiencing daylight, and which part is dark?

2. While making sure that the axis remains pointed at the same point on the ceiling, slowly spin the skewer or pencil that represents Earth's axis. You should be able to see Simon move through daylight and night. For what fraction of a 24-hour day does Simon experience daylight? Compare this with another pushpin person on the equator. How bright is the sunlight falling on Simon compared with the sunlight at the equator?

3. What effect does removing the tilt of the axis have? What would happen if Earth's axis were horizontal?

WINTER IN THE NORTHERN HEMISPHERE

1. Hold your foam ball so that it represents Earth at the winter solstice—the shortest day of the year in the Northern Hemisphere.

2. For what fraction of a 24-hour day does Simon experience daylight? How does this compare with what the pushpin person on the equator is experiencing? How bright is the sunlight that Simon is experiencing compared with the sunlight at the equator?

SAFETY PRECAUTIONS

1. Safety glasses or safety goggles are required for this activity.

2. Skewers and pushpins are sharps and can puncture or cut skin. Handle with care.

3. Lamps produce heat, which can burn skin. Handle with care.

4. Keep lamps away from any water source to prevent shock.

5. Make sure any fragile items or trip/fall hazards are removed from the work area.

PROBLEM 1

Teacher Guide

Northern Lights

Problem Context

The problem is posed from the point of view of a student who is moving from Iowa to Alaska. He is concerned that Alaska will be very cold. Students compare the seasonal temperatures, lengths of day, and the Sun's paths in Nashville (Tennessee), Sioux City (Iowa), and Fairbanks (Alaska) to find patterns that explain the seasonal changes in the Northern Hemisphere. The problem can be adjusted to compare any set of cities, including the students' hometown. However, be aware that large bodies of water or mountains may affect a town's seasonal weather changes.

Problem Solution

Model response: Earth is heated unevenly by the Sun. Because of the tilt of Earth's axis, the Northern Hemisphere gets more direct sunlight from March through September, but the Southern Hemisphere gets more direct sunlight from September through March. More direct sunlight is more concentrated. [CC 5: Energy and Matter: Flows, Cycles, and Conservation] In the Northern Hemisphere, not only is the sunlight more direct between the vernal equinox and the autumnal equinox, but the days are longer, too. The combination of length of daylight and directness of the Sun's rays explains the seasonal weather changes. Even though the sunlight is never very direct, Fairbanks is not frigid in the summer, because it experiences very long days. [CC 1: Patterns; CC 2: Cause and Effect: Mechanism and Explanation]

Activity Guide

The goal of this problem is for students to find patterns in data (SEP 4: Analyzing and Interpreting Data; CC 1: Patterns) on seasonal temperatures, lengths of daylight, and the angle of the Sun's rays and the Sun's path across the sky and then to explain the patterns using models (SEP 2: Developing and Using Models). We recommend that some of the data (times of sunrise and sunset and the Sun's path) be local. Students can mark the Sun's path by creating a sundial in the school yard by using chalk to mark the end of a stick's shadow at different times of day. To help students find patterns in their data, we encourage you to have students compile their data in tables (or graphs) that will organize them. Once students find the patterns, they should explain the patterns using models of the tilted

Earth. The Solar Motion Demonstrator (see the Extras page at *www.nsta.org/pbl-earth-space*) is a handheld model that students can make from paper that shows the path of the Sun from any latitude at any time of year. We recommend that they complement this activity, which shows seasonal changes from an earthbound perspective, with model-building exercises outlined on Page 3. These models allow students to step back and view the Earth-Sun system from multiple perspectives (SEP 2: Developing and Using Models).

You can adjust the degree of detail of the instructions for the modeling activity to fit your students' grade level. The goal is to develop their model so that they can use it to answer their own "need to know" questions. The instructions on Page 3 are meant to have an intermediate amount of information. You can define more of the steps or fewer. For example, you could also have students record what they see at the vernal equinox and the autumnal equinox. But, again, in the end you want them to use this system to answer their own questions.

TIPS FOR MODELING

When using the ball and light to model Earth and the Sun, students will need to explore the effects of the angle of the tilt of Earth's axis on length of daylight and light intensity. As they do this, they are likely to have difficulty keeping the axis pointing in the same direction, and you may want to have them establish an end of the room that the axis is always tilted toward. Students will be able to observe the day/night boundary on their foam balls more clearly if you have a darkened room to work in where the lamp (representing the Sun) is the only light.

Students can experience the effects of changing the directness of the Sun's rays by holding a flashlight 6 inches from a piece of paper. If they hold the flashlight perpendicular to the paper, the area of light from the flashlight will be smaller and more concentrated than if they hold the flashlight at an angle. If the flashlight has an incandescent bulb, students may be able to feel the difference in intensity if they substitute their hand for the paper. For a quantitative study, students can have a partner trace the shape of the light spot on the paper while they hold the flashlight, cut it out of the paper, and weigh it. The oblong spot from the flashlight held at an oblique angle will weigh more, indicating that it is larger and therefore has less light per unit area.

PROBLEM 1

Assessment

Northern Lights

Transfer Task

The table below shows the times of sunrise and sunset on the same day for three cities in the United States. Use the table to answer the following questions:

a. It is the time of year when temperatures are getting warmer. Approximately what time of year is it? Explain your answer.

b. Which city is farthest north? Which city is farthest south? Explain your answers.

CITY	TIME OF SUNRISE	TIME OF SUNSET
A	08:11:37	18:46:11
B	08:33:41	18:49:40
C	08:11:43	19:11:27

Model response:

a. All of the cities are experiencing less than 12 hours of daylight. This means that it is sometime between the autumnal equinox and the vernal equinox. Because the weather is getting warmer, it must be sometime between the winter solstice and the vernal equinox. [It is February 14.] [SEP 4: Analyzing and Interpreting Data; CC 1: Patterns]

b. City B [Fargo, North Dakota] is farthest north, and City C [Dallas, Texas] is farthest south. In the late winter, the city farthest north will have the shortest day, and the city farthest south will have the longest day. [CC 1: Patterns]

Application Questions

APPLICATION QUESTION 1

The 45th parallel runs through the Lower Peninsula of Michigan, Wisconsin, Minnesota, Idaho, Montana, and Oregon. It is halfway between the equator and the North Pole. In Figure 7.4, draw a picture of the Sun's path in the sky over the 45th parallel on the equinox. (The oval indicates the horizon. The 45th parallel runs through the middle of the oval on an east-west line.) [SEP 2: Developing and Using Models; CC 1: Patterns]

Figure 7.4. The Sun's Path Over the 45th Parallel

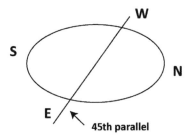

Model response: Because it is the equinox, we know that the Sun will rise due east and set due west. Because we are north of the Tropic of Cancer, we know that the Sun will never go directly overhead. It will always be to the south. [CC 1: Patterns; see the diagram (Figure 7.5)].

Figure 7.5. Diagram of the Sun's Path for Model Response

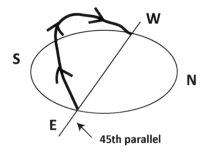

APPLICATION QUESTION 2

Your friend is moving from Michigan to Dallas, Texas. She is worried about how different it will be from what she is used to. Based on your understanding of seasons, the water cycle, and weather, answer the following questions for her. [SEP 6: Constructing Explanations and Designing Solutions; CC 1: Patterns; CC 5: Energy and Matter: Flows, Cycles, and Conservation]

a. Will the summer days be longer or shorter in Texas compared with Michigan? Explain your answer.

b. Will the winter days be cooler or warmer in Texas compared with Michigan? Explain your answer.

c. Much of the air that moves over Texas comes from the Gulf of Mexico. How will that influence the temperature and humidity of Texas? Explain your answer. (This question is related to Problem 2 on pages 159–171.)

Model response:

a. The summer days will be shorter in Texas, because the path of the Sun will be shorter and the Sun will move more directly overhead.

b. The winter days will be warmer in Texas, because Texas is farther south and therefore receives more direct sunlight and has longer winter days.

c. Air coming from the Gulf of Mexico will be warm and moist like the surface water below it.

[CC 1: Patterns; CC 2: Cause and Effect: Mechanism and Explanation]

Problem 2: Water So Old

Alignment With the *NGSS*

PERFORMANCE EXPECTATIONS	• *2-ESS2-3:* Obtain information to identify where water is found on Earth and that it can be solid or liquid. • *MS-ESS2-4:* Develop a model to describe the cycling of water through Earth's systems driven by energy from the Sun and the force of gravity.
SCIENCE AND ENGINEERING PRACTICES	• Developing and Using Models • Analyzing and Interpreting Data • Constructing Explanations and Designing Solutions • Obtaining, Evaluating, and Communicating Information

DISCIPLINARY CORE IDEAS	• *ESS2.C: The Roles of Water in Earth's Surface Processes*
	
	   ○ *Grades K–2:* Water is found in the ocean, rivers, lakes, and ponds. Water exists as solid ice and in liquid form.
	
	   ○ *Grades 6–8:* Water continually cycles among land, ocean, and atmosphere via transpiration, evaporation, condensation and crystallization, and precipitation, as well as downhill flows on land.
	
	   ○ *Grades 6–8:* Global movements of water and its changes in form are propelled by sunlight and gravity.

CROSSCUTTING CONCEPTS	• Patterns • Cause and Effect: Mechanism and Explanation • Energy and Matter: Flows, Cycles, and Conservation

Keywords and Concepts

Water cycle

Problem Overview

Two siblings debate whether the water they drink has always been on Earth and has been used by other living things.

Images available in full color on the Extras page (*www.nsta.org/pbl-earth-space*) are marked with the following icon: ☼.

PROBLEM 2

Page 1: The Story

Water So Old

Roberta and her younger brother Joe were sitting at the breakfast table. As usual, Roberta had her nose buried in her tablet.

"Joe, listen to this. It's the strangest idea!" said Roberta.

"Every idea you have is strange," replied Joe.

Roberta ignored his comment and began to read aloud.

"The water in your glass ..."

"How does that thing know I have a glass of water in front of me?" asked Joe.

Figure 7.6. Glass of Water

"Just be quiet and listen! The water in your glass may be 'fresh and clean,' but it has been around pretty much as long as Earth has! Some of the molecules of water in your glass may have run over the land before there was life on it. Maybe some were in the ancient oceans and were pushed by a jellyfish." (See Figures 7.6 and 7.7.) Roberta looked up and added, "I'll bet some of your water was in the swamps that were full of insects and ferns and palms before the dinosaurs. Maybe some rained down on the dinosaurs that came later and they splashed through it."

✪ **Figure 7.7. Jellyfish**

"Yuck! My water has had animal feet in it?" said Joe, frowning.

"Maybe a Neanderthal drank some of your water," continued Roberta.

"What's a Neanderthal?" asked Joe.

"Can't you pay attention for one second, Joe? See that apple you are eating? The water in that apple might have been in a cloud in the sky over our great-great-great-grandfather, or maybe it was in the glaciers that shaped Michigan."

"I think it just came out of the town's well," retorted Joe. Joe suspected that his sister had found an online article with fake information, and he would like to prove his know-it-all sister wrong.

Your Challenge: *Help Joe outdo his sister. His sister believes that water has been around pretty much since Earth began, that it is present in many forms, and that it simply moves and changes forms. Either give Joe evidence to prove his sister wrong or help him explain to his sister why these ideas are not so strange.*

PROBLEM 2

Page 2: More Information

Water So Old

Roberta mentioned the water in the apple that Joe was eating (see Figure 7.8) and in Joe's glass falling as rain or running over the land, being in the ocean, in glaciers, in swamps, or in clouds, or even being in other living things. Joe wondered if water could be found in any other types of places. Roberta's article implied that water moved around between all of these types of places. How could that happen? Joe knew that rivers flowed into the oceans, but could water ever get out of the oceans?

And Roberta's article made it sound like water had been moving around in the same ways forever or at least for a very long time. But wasn't Earth different in the time before the dinosaurs? (See Figure 7.9.) Was there anything making water or turning water into something else?

✪ **Figure 7.8. Apple**

✪ **Figure 7.9. Dinosaur**

Your Challenge: *Help Joe outdo his sister. His sister believes that water has been around pretty much since Earth began, that it is present in many forms, and that it simply moves and changes forms. Either give Joe evidence to prove his sister wrong or help him explain to his sister why these ideas are not so strange.*

PROBLEM 2

Page 3: Resources and Investigation

Water So Old

Resources

- A video of the water cycle is available from the National Aeronautics and Space Administration (NASA) at *www.youtube.com/watch?v=0_c0ZzZfC8c.*

- An interactive diagram of the water cycle is available at *http://water.usgs.gov/edu/watercycle.html.*

Water Cycle Model Investigation

MATERIALS

- Tap water
- Ice cubes
- Lamp (preferably with a 100 W incandescent bulb)
- Beaker
- Baking dish or cake pan
- Soil, sand, and pea gravel
- Modeling clay
- Aquarium tank or other clear container with a lid
- Spray bottle or watering can

PROCEDURES

INVESTIGATING WATER VAPOR

1. Place a shallow dish of hot water near one end of a fish tank with a lid. Tightly affix the lid to the fish tank. Set up a lamp so that it shines on the water. Allow the tank to sit for an hour or more.

2. Place a beaker or cake pan of ice on the cover at the other end of the tank. What do you observe? What part(s) of the water cycle does this represent? Is this evidence in support of or against Joe's sister's ideas?

3. How could you use this model to investigate the effects of different variables on how water changes and moves?

INVESTIGATING GROUNDWATER

1. To build a model of groundwater (see Figure 7.10), spread a thin layer of modeling clay on the bottom of an aquarium or other clear container. This represents a confining layer of bedrock. Cover this with a few inches of pea gravel, then a few inches of sand, and finally some soil.

2. To represent a lake or river, leave one end of the container covered only with clay. Gently pour water onto your lake (or river) until it stops soaking into the ground and about half an inch remains. (See Figure 7.10.)

3. See what happens if it rains on the land (but not on the lake). You can use a spray bottle or a watering can to simulate rain. What part of the water cycle does this represent? Is this evidence in support of or against Joe's sister's ideas?

4. How could you use this model to investigate the effects of different variables on how water changes and moves?

Figure 7.10. Groundwater Model

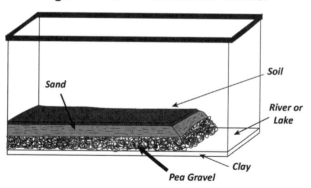

SAFETY PRECAUTIONS

1. Use caution when working with heat sources—they can burn skin!

2. Use only GFI-protected circuits when working with electrical equipment, to prevent shock.

3. Be careful not to splash water on a hot bulb. It can shatter the bulb, creating a sharp and projectile hazard.

4. Review safety data sheet (SDS) for clay.

5. Clean up clay before it dries to prevent respiratory exposure.

6. Immediately wipe up any water spilled on the floor to avoid a slip or fall hazard.

7. Wash hands with soap and water when the activity is complete.

PROBLEM 2

Teacher Guide

Water So Old

Problem Context

This problem focuses on the constant cycling of water through many forms and reservoirs. The initial story suggests many places and times where water exists. These can be altered to include any experiences students are likely to have had.

Problem Solution

Model response: The movements and changes of water as it cycles are driven by gravity and sunlight heating Earth. Gravity moves liquid water and ice in glaciers downhill. It also pulls precipitation (rain or snow) down from the sky to the surface of Earth. When sunlight heats water by shining on it directly or heating the air, the water may change from solid to liquid (melt) or to gas (sublimate) or change from liquid to gas (evaporate). When water cools, releasing heat into the surroundings, it may change from gas/vapor to liquid (condense) or from liquid to solid (freeze). Evaporation is the main way in which water goes from low to high places. Within the ocean, convection currents also caused by heating and cooling cause water to move around. [CC 5: Energy and Matter: Flows, Cycles, and Conservation]

Precipitation (water falling from the sky in liquid or solid form) is "fresh," not salty. It contains very little (if anything) besides water and dissolved carbon dioxide. As it flows over and through the earth, it may carry other substances with it. Some of these it brings to the ocean [see Figure 7.11] where they accumulate, thus creating the saltwater. The water in the ocean can evaporate and go elsewhere, but the salts and other contaminants are stuck in the oceans. Slowly, over time, they form sediments on the ocean floor.

All the water on Earth is part of this cycle. New water is not made, nor is water turned into anything else. There is only one exception to this, and that is for water entering a living organism. The water may go through the organism without being changed. Plants move a lot of water from the ground to the atmosphere (transpiration). But plants may also change the water. The water may react with carbon dioxide (photosynthesis) and become plant material and potentially part of other organisms that eat the plant. But organisms are eventually recycled, and water will be reformed (respiration). [CC 4: Systems and System Models; CC 5: Energy and Matter: Flows, Cycles, and Conservation]

✪ Figure 7.11. Ocean

Activity Guide

Students should be able to use the story and their own experiences, especially with weather, to identify many forms of water and processes that move and change water. The NASA video referenced on the Resources page shows an animation of the many processes of the water cycle without narration and can be used to help students fill in processes.

Much of the water cycle is visible and familiar to students. However, two forms are not: water vapor and groundwater. Students can investigate water vapor by trying to get liquid water to evaporate and condense in a closed system. They can put some water in a closed, clear container in sunlight. After some time, condensation will form on the sides or top of the container. This is water that has evaporated and condensed in another place. Alternatively, students can make the cycling of water happen more quickly by placing a shallow dish of water at one end of a closed, clear container such as a fish tank with a lid. The investigation in the Page 3: Resources section of this problem describes that model in more detail. [*Note:* Under the right conditions, students may see a cloud form near the ice, and they will probably see condensation form on the sides and lid. If there is enough water vapor, it may even "rain"!] Students can experiment to see the effect of changing different variables. Variables might include putting dark construction paper on one side of the tank; using boiling water or cold water; or changing the amount of ice, the size of the tank, or even the color of the light.

The other part of the water cycle that is not visible is groundwater, and students are apt to overlook the connection between groundwater and surface water. Students can model groundwater and a stream or lake in an aquarium tank or other clear container. The second part of the investigation in the Resources section describes a possible activity relating to groundwater. Students can experiment with changing the variables to see their effects. They may choose to experiment with ever-increasing amounts of rain on their system or with different shapes of streambeds, tilting the tank to simulate a slope or adding cocoa mix to the water to simulate silt or pollution in the water. Students can also experiment with methods to control erosion in their model.

The class can use their models to answer their "need to know" questions. They should discuss how their models compare to natural systems and what evidence they have gathered that they can use to help Joe.

The National Science Foundation has a video, available at *www.youtube.com/watch?v=al-do-HGuIk*, that explains all of the processes that move and change water in the water cycle. You can use the video to summarize the water cycle after students have come up with their own ideas.

SAFETY PRECAUTIONS

1. Indirectly vented chemical-splash goggles, aprons, and nonlatex gloves are required for this activity.

2. Immediately wipe up any water spilled on the floor to avoid a slip or fall hazard.

3. Wash hands with soap and water when the activity is complete.

PROBLEM 2

Assessment

Water So Old

Transfer Tasks

TRANSFER TASK 1

Water cycling is illustrated in Figure 7.12; each arrow represents water moving or changing. Describe the processes represented by each arrow by filling in the table (below the figure). Based on the information in your table, explain how this cycle redistributes energy in the system.

Figure 7.12. Simplified Water Cycle Diagram

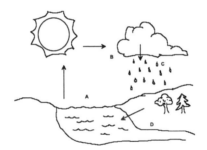

ARROW	NAME OF PROCESS	PHASE CHANGE (IF ANY)	DRIVING FORCE OR ENERGY TRANSFER
A			
B			
C			
D			

Model response:

ARROW	NAME OF PROCESS	PHASE CHANGE (IF ANY)	DRIVING FORCE OR ENERGY TRANSFER
A	Evaporation	Liquid to gas	Absorbing heat from system
B	Condensation	Gas to liquid	Releasing heat to the system
C	Precipitation— combining of droplets to form drops that fall	None	Attraction between water molecules Gravity
D	Runoff or groundwater recharge	None	Gravity

This cycle takes thermal energy/heat from the lake or river and releases it into the atmosphere. The energy needed for the water to evaporate comes from the lake or river and is released into the atmosphere when the water condenses to form a cloud. [CC 4: Systems and System Models; CC 5: Energy and Matter: Flows, Cycles, and Conservation]

TRANSFER TASK 2

Write a story following a water molecule from its time in the primordial oceans before there was life on land to Joe's glass of water. Your story should include at least four of the places that Roberta and Joe talked about. You should also note when the water molecule was part of a process that transferred energy. [CC 5: Energy and Matter: Flows, Cycles, and Conservation]

Model response [*Note to teacher:* Answers will vary with students' choices of places for the water molecule to visit.]: My goodness, I seem to have been in this ocean for a REALLY long time. It's been ages since anything happened. The last exciting event that I remember was a thousand years ago when that jellyfish gave me quite a downward push. It's too bad my friend, Gracie, evaporated last week. I guess I shouldn't complain. It is nice and warm here at the surface and I have been going faster. Oh my, what's this? I'm evaporating! That last bit of energy from my neighbors set me free! Up, up, and away!

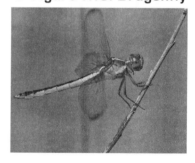

✪ **Figure 7.13. Dragonfly**

Some time (eons) later: This time around, it seems like I've been here in the atmosphere for a long time. I'd like to find some other molecules so that we could condense and precipitate. Last time I was on land, things were changing, and I would like to have another look. Hello, over there! Would you like to join me in a droplet? We're slowing down anyway. Oh good, you would? It's nice to be together with someone again and to get rid of all of that energy. And there are some more folks. Yes, I think we're big enough to fall now that we've all gotten together. Here we goooooooo! My, it's lovely and green down in this swamp and comfortably warm. Look at all of those huge dragonflies zooming overhead. [See Figure 7.13.] Gosh, I'm speeding up again already. Too much sunlight, I guess. And there's that extra energy that sends me off on my own again. Good-byeeeeeee!

Some time (eons) later: I've always been told that we water molecules spend most of our time in the ocean, but I seem to be the one that evaporates quickly back into the atmosphere. And here I am again. There seems to be a big updraft here and it's getting cold quickly as we go up. I am slooooowing down. Oh, hello there. What a beautiful snowflake you all make. Sure, I'd like to join. I didn't want all that energy anyway. Much better to drift down with you all. And we've touched down already. I don't think I've ever seen so many flakes. Yikes, we're piling up quickly. I've never been in a glacier before. I'm feeling squished. Maybe the best thing to do is go to sleep. We are moving downhill so very slowly, I don't think anything is going to happen for a very long time.

Some time (eons) later: Back in the atmosphere again! But this looks like fun. I'm over farmland. Lots going on. You can always spot the farmer. This one is working very hard hoeing, milking the cows, feeding the pigs and chickens, mucking out the barn. I hope his great-great-great-grandchildren appreciate how hard he works. His corn seems to be doing very well. I keep running into other water molecules that have transpired from the corn.

Some time (eons) later: Gosh, I thought my time in the glacier was boring. Being in this aquifer is no fun either. I can move past my colleagues, but nothing much happens. Oops, I spoke too soon. This must be the well that someone was telling me about. I like the smooth pipes. And now I'm in a clear cylinder. What a wonderful world … Oops, someone's drinking us. I guess we're in for more dark.

Application Question

Explain how a water molecule could get from the bottom of the ocean to the top of Mount Everest. Your explanation should include what drives the water molecule's movements and changes. [SEP 2: Developing and Using Models; SEP 6: Constructing Explanations and Designing Solutions; CC 5: Energy and Matter: Flows, Cycles, and Conservation]

Model response: The water molecule would have to be carried by a current to the surface of the ocean. The currents are driven by uneven heating of the ocean. Once on the surface, the water molecule could evaporate if it absorbed enough energy from sunlight and the surrounding air and water. It would become part of a warm, moist air mass. Then it would have to be lifted up. This could happen if its air mass converged with a cooler, drier air mass. The denser air would push under warm, moist air. The Himalayas would also deflect the air upward, where it would cool. As the water molecule cooled, it would condense, bonding to other water molecules and forming a cloud and eventually a snowflake that could fall on Mount Everest. [CC 5: Energy and Matter: Flows, Cycles, and Conservation]

Problem 3: Leave It to the Masses

Alignment With the *NGSS*

PERFORMANCE EXPECTATIONS	• *MS-ESS2-5:* Collect data to provide evidence for how the motions and complex interactions of air masses result in changes in weather conditions. • *MS-ESS2-6:* Develop and use a model to describe how unequal heating and rotation of Earth cause patterns of atmospheric and oceanic circulation that determine regional climates.
SCIENCE AND ENGINEERING PRACTICES	• Developing and Using Models • Analyzing and Interpreting Data • Constructing Explanations and Designing Solutions • Obtaining, Evaluating, and Communicating Information

DISCIPLINARY CORE IDEAS	• *ESS2.C: The Roles of Water in Earth's Surface Processes*
	
	  ○ *Grades 6–8:* The complex patterns of the changes and the movement of water in the atmosphere, determined by winds, landforms, and ocean temperatures and currents, are major determinants of local weather patterns.
	
	  ○ *Grades 6–8:* Global movements of water and its changes in form are propelled by sunlight and gravity.
	
	• *ESS2.D: Weather and Climate*
	
	  ○ *Grades 6–8:* Weather and climate are influenced by interactions involving sunlight, the ocean, the atmosphere, ice, landforms, and living things. These interactions vary with latitude, altitude, and local and regional geography, all of which can affect oceanic and atmospheric flow patterns.
	
	  ○ *Grades 6–8:* Because these patterns are so complex, weather can only be predicted probabilistically.

CROSSCUTTING CONCEPTS	• Patterns • Cause and Effect: Mechanism and Explanation • Systems and System Models • Energy and Matter: Flows, Cycles, and Conservation

Keywords and Concepts

Air masses, weather, and weather forecasting

Problem Overview

Students in a class are invited to help report the weather on TV and need to learn about air masses and how they help predict the weather.

Images available in full color on the Extras page (*www.nsta.org/pbl-earth-space*) are marked with the following icon: ⊙.

PROBLEM 3

Page 1: The Story

Leave It to the Masses

Every Thursday evening, your local TV station, WWET, has a group of students give the weather forecast on the 6 o'clock news. Your class has been chosen for next Thursday. Ray Nye Day, the regular meteorologist, sent you the following information and advice on how to prepare:

Before you give the local forecast, you will show four weather maps of the United States: temperature, wind, Doppler radar, and a classic weather map with highs and lows marked. That sounds like a lot, but they are all related. If you understand air masses, you will be able to explain any of the maps. So my advice to you is to learn about air masses. Here are some critical ideas about air masses:

- An *air mass* is a large body of air with similar temperature and moisture properties throughout.

- If an air mass stays in place for very long, it takes on the characteristics of the surface below.

- Warmer air is less dense than cooler air because the molecules are moving faster, colliding harder, and spreading out. Therefore, cooler air has higher pressure than warmer air.

- Drier air is more dense than wetter air because nitrogen and gas molecules (the most abundant molecules in air) are heavier than water molecules. Therefore, dry air is associated with higher pressure than wet air.

- *Fronts* are boundaries between air masses. These are often places where the weather is changing.

Your Challenge: *Prepare for your TV appearance by learning about weather fronts. You will know you are ready if you can identify air masses on the weather maps and if you can explain the weather that is happening at the fronts or the source of any storms.*

PROBLEM 3

Page 2: More Information

Leave It to the Masses

Mr. Day has a set of weather maps for you to study and practice on (see Figures 7.14–7.17). They show the temperature, wind, radar, and fronts for one day. He suggests that you look for patterns across the different types of maps. He says that there is a lot of information in these maps but that the weather in these maps is typical for this time of year.

Figure 7.14. Temperature Map for March 23, 2016

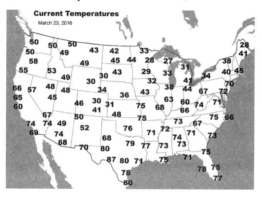

Figure 7.15. Wind Map for March 23, 2016

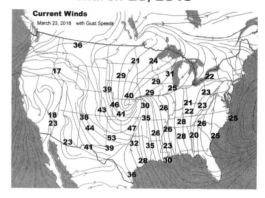

Figure 7.16. Doppler Radar Map for March 23, 2016

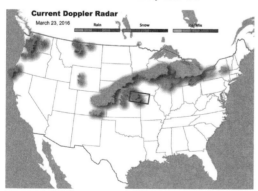

Figure 7.17. Weather Map for March 23, 2016

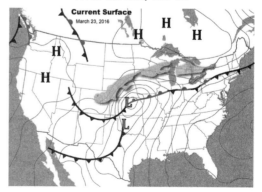

Mr. Day also sent you two diagrams of different types of fronts (see Figure 7.18). He said they should help when you are learning about fronts.

✪ Figure 7.18. Other Weather Resources From Mr. Day

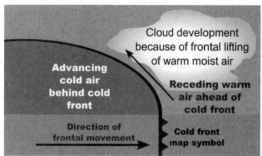
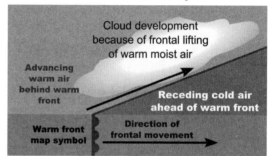

Your Challenge: *Prepare for your TV appearance by learning about weather fronts. You will know you are ready if you can identify air masses on the weather maps and if you can explain the weather that is happening at the fronts or the source of any storms.*

PROBLEM 3

Page 3: Resources

Leave It to the Masses

Online Resources on Weather

- Current weather maps of any type are available at *http://weather.com/maps*.

- This animated website shows current wind conditions and includes a section that explains how data are used to generate the animation: *http://hint.fm/wind*.

- Current and archived weather data plus information about clouds and precipitation, hurricanes, and El Niño weather patterns are available at *http://ww2010.atmos.uiuc.edu/%28Gh%29/home.rxml*.

- To explore a variety of different weather fronts and the interactions of air masses, view the simulator at *www.phschool.com/atschool/phsciexp/active_art/weather_fronts*.

Meteorology Books

Ahrens, C. D. 2016. *Meteorology today: An introduction to weather, climate, and the environment.* 11th ed. Boston: Cengage Learning.

Cox, J. D. 2000. *Weather for dummies.* New York: Hungry Minds.

Williams, J. 1997. *The weather book: An easy-to-understand guide to the USA's weather.* New York: Vintage.

PROBLEM 3

Teacher Guide

Leave It to the Masses

Problem Context

In this problem, students prepare to do a weather forecast on their local TV channel, with the same weather maps used by the meteorologist.

Problem Solution

Model response: Air masses are evident on the temperature and wind maps as abrupt changes. On the maps Mr. Day gave us, the temperatures of adjacent air masses differ by as much as 15° in temperature. On the wind map, winds from the north or northwest meet winds from the south. The air mass coming in from the north is cold (see temperature map) and probably dry, because it's coming from Canada. The air mass coming from the south is warm (see temperature map) and probably moist, because it's coming from the Gulf of Mexico. On the classic weather map, the boundary between air masses where one air mass is pushing into another is shown with red or blue lines (sometimes black). [SEP 4: Analyzing and Interpreting Data] These boundaries are called fronts, and they're where most of the precipitation occurs. In this case, the cold front is denser than the warm front, so it is deflecting the warm front upward. As the warm, moist air rises, it cools, condenses, and forms raindrops. The rain is visible on the radar map. [CC 2: Cause and Effect: Mechanism and Explanation]

The classic weather map (the name used by Weather.com for maps showing highs and lows) also shows the areas of high (H) and low (L) pressure. In areas of high pressure, the air is dense and pushes down and the winds spiral out in a clockwise rotation. In areas of low pressure, the air is less dense. Denser air from outside the low-pressure area spirals into the area with a counterclockwise rotation.

Activity Guide

The first step in finding patterns in the different types of weather maps is to understand how the information is represented. (Comparing the animations on *http://hint.fm/wind* to current wind maps may help students better understand wind maps.) Then students can try drawing in air mass boundaries by looking for abrupt changes across a map. They can compare these across map types for a particular day. It is not unusual for these boundaries to differ slightly between map types; however, there should be overall agreement. The temperature of the air masses will be evident from the temperature map, but students will have to hypothesize about the relative humidity based on each air mass's origins. They should then be able to use the pictures and descriptions of the fronts to explain any precipitation.

PROBLEM 3

Assessment

Leave It to the Masses

Transfer Task

In the winter when cold winds come from the west and move across Lake Ontario, the western edge of New York gets a lot of lake-effect snow (see Figure 7.19). This phenomenon is a miniature version of a warm, moist air mass hitting a colder, drier one. [CC 1: Patterns] Using Figure 7.19 below, explain why Buffalo, New York, experiences greater lake-effect snows than Albany, New York. In your explanation, discuss phase changes in water and air density. [SEP 6: Constructing Explanations and Designing Solutions].

Figure 7.19. Lake-Effect Snow in New York

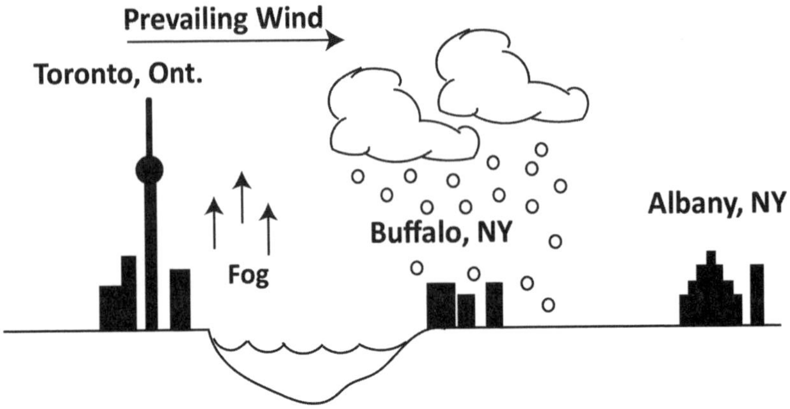

Model response: Lake Ontario is a large body of water that cools more slowly than the surrounding air. As the prevailing west wind moves over the lake, water evaporates from the warmer water, and the air picks up water vapor. [CC 1: Patterns] When the warmer, moist air encounters the cooler, denser air over New York, it is pushed up. This causes the air to cool and the moisture in it to form snow. Most of the moisture in the air falls very near the coast, and there is little left when the air reaches Albany. [CC 2: Cause and Effect: Mechanism and Explanation]

Application Question

Use a current set of maps (Doppler radar, temperature, wind, and classic) to predict tomorrow's weather for your hometown. Explain your prediction using information from the maps. (Current weather maps of any type are available at *http://weather.com/maps*.)

> *Model response:* Responses will depend on current conditions. Students should use the maps to determine the air masses over the United States. [SEP 4: Analyzing and Interpreting Data] By looking at the air mass (its temperature and humidity) currently over their town and the direction and speed at which that mass is moving, they should be able to determine if conditions for the next day will change significantly. They should expect precipitation where air masses will collide, unless both masses are dry. [SEP 6: Constructing Explanations and Designing Solutions]

Weather Problems: General Assessment

General Questions

GENERAL QUESTION 1

Figure 7.20 shows the Sun's rays falling directly on the Tropic of Capricorn in the Southern Hemisphere (23.5° S). On this day at noon, the Sun is directly overhead if you are standing on the tropic. What day of the year is it? Explain your answer.

Figure 7.20. The Sun's Rays Falling on Earth

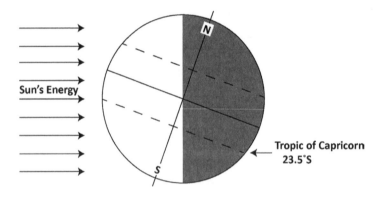

Model response: It is December 21. The direct rays of the Sun are falling at the far southern end of their range. It is the beginning of summer in the Southern Hemisphere. [SEP 2: Developing and Using Models; CC 1: Patterns]

GENERAL QUESTION 2

Explain why water does not disappear from the water cycle when an animal drinks water.

Model response: When the animal drinks water, the molecules of water are absorbed into its body. They will be returned to the atmosphere when it exhales, sweats or pants, or excretes them. If they remain in its body until the animal dies, they will be released back into the water cycle when its body decomposes. [SEP 6: Constructing Explanations and Designing Solutions; CC 5: Energy and Matter: Flows, Cycles, and Conservation]

GENERAL QUESTION 3

Explain why when a cool, dry air mass moves into a warm, moist air mass, there is likely to be rain at the front.

> *Model response:* The cool, dry air mass is denser than the warm, moist air mass. Because it is cooler, the air molecules are moving more slowly and are less spread out. The denser cool, dry air pushes under the less dense, warm, moist air. The warm, moist air is displaced upward. As it rises, it cools. The water vapor condenses, forming droplets that are visible as clouds. If there is enough moisture, the droplets will form drops big enough and heavy enough to fall as rain. [SEP 6: Constructing Explanations and Designing Solutions; CC 2: Cause and Effect: Mechanism and Explanation]

Common Beliefs

Indicate whether the statements are true (T) or false (F), and explain why you think so (*model responses shown in italics*).

1. The Sun's light is very steady and uniform, but Earth is not heated uniformly by it. *(T) Not all of the Sun's light reaches the surface of Earth. Some of it is reflected by clouds. When sunlight does reach Earth's surface, some of it may be reflected without being absorbed if the surface is light in color (e.g., snow covered). Depending on the relative position of the Sun and Earth, the sunlight will fall more or less directly on Earth's surface. All of these factors result in uneven heating of Earth.*

2. Seasons are caused by differences in Earth's distance from the Sun. *(F) We have seasons because the tilt of Earth's axis causes different latitudes to receive light of different durations and intensities.*

3. In the Northern Hemisphere, the farther north you go, the longer the summer days are. *(T) In the Northern Hemisphere, as you go north, the path of the Sun in the sky changes. The Sun does not go as high above the horizon, but it takes a longer path across the sky. It rises and sets from points farther north of an imaginary line running east and west and makes a lower arc in the sky.*

4. A waterfall that keeps flowing through the hot, dry summer months must have rain falling uphill from it. *(F) As with any stream or river, rain is only one of many ways that it is fed water. There might be melting snow upriver. This is particularly true downstream from mountains. Saturated ground may be discharging into the river.*

5. Water redistributes energy as it cycles. *(T) Water molecules that have melted or evaporated have acquired the energy it takes to break the intermolecular bonds/forces that hold them in a solid or liquid, respectively. If they move to a new place (downhill in surface water in the case of the liquid water*

or into the atmosphere in the case of water vapor), they take that energy with them. They release that energy when they turn back into their earlier state.

6. If upper-level and surface air pressure, temperature, and wind direction data are collected, a meteorologist can predict the weather, but these predictions are not always accurate. *(T) Because of the large number of factors that affect weather, it is very difficult to predict accurately for more than a few days out.*

7. When two air masses meet, clouds and rain develop along the front. *(F) Clouds and rain develop along fronts when at least one of the air masses contains moist air. However, if two air masses that differ in temperature but have little humidity meet, it is not likely that there will be rain.*

8

ASTRONOMY

The astronomy problems ask participants to explain the phases of the Moon, solar and lunar eclipses, and the apparent motion of planets. The problems address both descriptions and explanations of observations of the night sky.

The astronomy problems focus on the crosscutting concepts of Patterns (CC 1) and Scale, Proportion, and Quantity (CC 3) described in the *Next Generation Science Standards* (*NGSS*; NGSS Lead States 2013). The periodic motions of celestial objects explain why we see predictable patterns in the observations from Earth, and the relative position and size of the Moon, the Sun, and Earth make eclipses possible. This chapter engages students in a number of *NGSS* science and engineering practices (SEPs): Developing and Using Models (SEP 2); Analyzing and Interpreting Data (SEP 4); Constructing Explanations and Designing Solutions (SEP 6); Engaging in Argument From Evidence (SEP 7); and Obtaining, Evaluating, and Communicating Information (SEP 8). Disciplinary core ideas (DCIs) addressed in this chapter are The Universe and Its Stars (ESS1.A) and Earth and the Solar System (ESS1.B).

Big Ideas
Phases of the Moon

- The phases of the Moon (the Moon's appearance, position in the sky, and rise and set times) change in a regular and predictable pattern.

- All observers on Earth see nearly the same phase of the Moon on any given day or night.

- The Moon takes about a month to orbit Earth and go through a cycle of phases.

- The Moon appears to rise in the east and set in the west as Earth rotates.

- The Moon shines by reflected sunlight.

- Observers on Earth only see the part of the Moon that is illuminated by the Sun *and* that is facing Earth.

- The phases of the Moon change because the Moon is revolving around Earth.

Solar and Lunar Eclipses

- During a lunar eclipse, the entire hemisphere of Earth facing the Moon can watch the eclipse.

- During a lunar eclipse, totality can last an hour or more.

- A lunar eclipse occurs when the full Moon moves into the shadow of Earth.

- A solar eclipse occurs when the Moon moves in front of the Sun (new Moon) and blocks all or part of its light, casting a shadow on Earth.

- During a solar eclipse, totality lasts only a few minutes.

- During a solar eclipse, only those directly under the very narrow umbra part of the Moon's shadow will see the total solar eclipse.

- We do not have eclipses every month because the Moon's orbit is not aligned in the same plane as Earth's orbit.

- Both a solar and a lunar eclipse occur approximately every six months, when the new Moon and full Moon, respectively, are aligned with the Sun and Earth.

Motion of the Planets

- The celestial sphere and all the objects in it appear to move across the sky over the course of a day or night because Earth is rotating on its axis.

- Planets appear to drift slowly across the background stars over the course of days, weeks, months, or years because the planets (including Earth) are also orbiting the Sun.

- The location of a planet in the sky is determined by a planet's position in its orbit relative to Earth at any specific time. The speed of a planet's apparent motion in the sky is primarily a function of its orbit and Earth's orbit, because planets orbit the Sun at different speeds.

- The phase of a planet is determined by its orbit and its relative position in its orbit with respect to the Sun and Earth.

186   NATIONAL SCIENCE TEACHERS ASSOCIATION

Conceptual Barriers
Common Problems in Understanding

One of the challenges in learning about the phases of the Moon is the need to make a three-dimensional mental image of a very large system. The image needs to consider reflected light and multiple objects moving in space from the perspective of an observer on the surface of one of those objects … while that object is rotating and moving. This is a challenging task.

- Students often have difficulty explaining or picturing the motions of the objects involved from a point somewhere above the North Pole (a common perspective for drawings of the solar system) and understanding the implications of these motions for what we see from Earth.

- Some learners find it challenging to connect the Earth view to the space view of the arrangement of the Sun, Earth, or other objects.

- When the Sun is drawn as a nearby sphere radiating outward in all directions in a space-view diagram, it is difficult to deduce the shape of the Moon (crescent, quarter, gibbous, etc.) phases accurately (see Figure 8.1). A more accurate and less confusing diagram is to show the Sun's rays as parallel beams of light (see Figure 8.2).

Figure 8.1. Misconception About Moon Phases

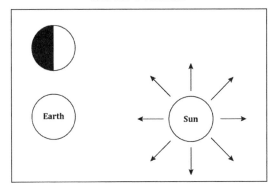

In this space-view diagram, the illumination of the first quarter Moon does not match the sunlight's angle of incidence to the Moon.

Figure 8.2. A More Accurate View of Moon Phases

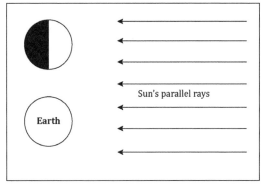

In this space-view diagram, the illumination of the first quarter Moon agrees with the angle of incidence of sunlight.

Figure 8.3. Solar Eclipse, Out of Scale

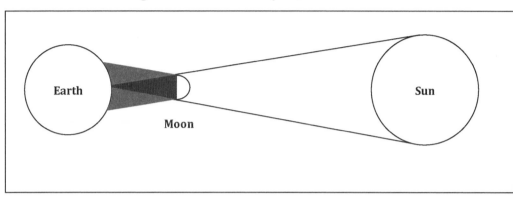

In this diagram, Earth and the Moon are the correct relative sizes but the Sun is not. Furthermore, the distances are not to scale. At these relative scales, the shadows are not their proper sizes and the Moon's orbital motion is highly exaggerated.

Figure 8.4. Solar Eclipse, in Scale

In this diagram, the relative sizes of Earth and the Moon and their distance are shown correctly. The Sun cannot be shown to correct size or distance. For clarity, only the umbra is shown.

- Another common misrepresentation relates to the size of the Moon's umbra on Earth's surface during an eclipse. To fit diagrams on a page, the distances between the Sun, the Moon, and Earth are shown on a different scale than the relative sizes (see Figure 8.3). This leads to inaccurate depictions of the area on Earth that will see an eclipse. A more accurate diagram is shown in Figure 8.4.

Common Misconceptions

This section includes a compilation of common misconceptions about the Moon and about planets. Many of these misconceptions arise as students substitute simpler explanations based on familiar experiences. For example, many objects in the night sky, such as stars, shine with their own light; therefore, some students assume that the Moon shines with its own light. Sometimes when we can't see the Moon or the Sun, it is because clouds are blocking our view. Therefore, some students assume that phases of the Moon are caused by clouds.

These misconceptions have two important implications for teaching astronomy. First, very few students have made careful observations of the day or night sky. Such observations would contradict many of the misconceptions. Second, the actual explanation is more complicated than the misconceptions, requiring that students be able to picture the relative position of three objects from different perspectives. The challenge of viewing the Moon's phases from multiple perspectives indicates the need to include extensive work with models in instruction.

COMMON MISCONCEPTIONS ABOUT THE MOON

- The Moon shines by its own light.

- The Moon is only "out" (visible) at night.

- Phases of the Moon are caused by shadows—the shadow of Earth (or another solar system object) on the Moon's surface.

- Phases of the Moon are caused by clouds.

- The Moon appears in the same place in the sky every night.

- The Moon rises and sets at the same time every night.

- People at different locations (latitudes, longitudes, or altitudes) on Earth see different phases of the Moon on the same day or night.

- The phases of the Moon correspond to a calendar month; that is, if the Moon is full June 17, it will be full again July 17 and every month on the 17th day of the month.

COMMON MISCONCEPTIONS ABOUT ECLIPSES

- Eclipses occur every month.

- Eclipses occur very rarely.

- Eclipses repeat on the same day every calendar year.

- Eclipses are always total.

- The Moon is full during a solar eclipse.

- The Moon is new during a lunar eclipse.

- The Moon disappears from view during a lunar eclipse.

- It is dangerous to be outside during a solar eclipse.

Box 8.1. Sample Interdisciplinary Connections for Astronomy Problems

- **Language arts:** Write a story about traveling through the solar system or visiting the International Space Station during an eclipse.

- **Geography:** Use online mapping tools to map the countries and cities that will be able to see a solar eclipse on a given date.

- **Social studies:** Read about the government's role in the oversight of organizations such as NASA and the National Oceanic and Atmospheric Administration (NOAA) that collect data about Earth and the solar system.

- **Mathematics:** Build a scale model of the Moon and Earth by calculating accurate distances between them, including a measurement that would describe the relative size of the Sun and its distance from Earth in your model.

- **Art and music:** Create a painting, drawing, or sculpture depicting the phases of the Moon. Search for and compare songs from the 20th century about the Moon.

COMMON MISCONCEPTIONS ABOUT THE PLANETS

- If a planet takes so many days or years to orbit the Sun, if we see it in a certain place in the sky now, it will be in the same position one planet's orbital period later.

- Since Earth takes one year to orbit the Sun, if we see a planet in a certain place in the sky, we will see it again in the same place next year at this time.

- The planets can be seen on any night.

- Planets are seen in the same places in the sky at night.

- Planets do not have phases; only the Moon has phases.

Interdisciplinary Connections

There are a number of interdisciplinary connections that can be made with the problems in this chapter. Connections can be made to language arts, geography, social studies, mathematics, and art and music (see Box 8.1).

Reference

NGSS Lead States. 2013. *Next Generation Science Standards: For states, by states.* Washington, DC: National Academies Press. *www.nextgenscience. org/next-generation-science-standards.*

Problem 1: E.T. the Extra-Terrestrial

Alignment With the *NGSS*

PERFORMANCE EXPECTATIONS	• *MS-ESS1-1:* Develop and use a model of the Earth-Sun-Moon system to describe the cyclic patterns of lunar phases, eclipses of the Sun and Moon, and seasons.
SCIENCE AND ENGINEERING PRACTICES	• Developing and Using Models • Analyzing and Interpreting Data • Using Mathematics and Computational Thinking • Constructing Explanations and Designing Solutions • Obtaining, Evaluating, and Communicating Information

DISCIPLINARY CORE IDEAS	• *ESS1.A: The Universe and Its Stars*
	   ◦ *Grades 6–8:* Patterns of the apparent motion of the Sun, the Moon, and stars in the sky can be observed, described, predicted, and explained with models.
	• *ESS1.B: Earth and the Solar System*
	   ◦ *Grades 6–8:* The solar system consists of the Sun and a collection of objects, including planets, their Moons, and asteroids that are held in orbit around the Sun by its gravitational pull on them. This model of the solar system can explain tides, eclipses of the Sun and the Moon, and the motion of the planets in the sky relative to the stars.

CROSSCUTTING CONCEPTS	• Patterns • Systems and System Models

Keywords and Concepts

Moon phases

Problem Overview

Students examine scenes from the movie *E.T.* to see if the phases of the Moon match the time frame of the story.

Images available in full color on the Extras page (*www.nsta.org/pbl-earth-space*) are marked with the following icon: ✪.

PROBLEM 1

Page 1: The Story

E.T. the Extra-Terrestrial

Citizens for Scientific Integrity (CSI) is a watchdog group that checks the scientific accuracy of movies and videos. The group has filed a suit against director Steven Spielberg. The suit alleges that he grossly misrepresented the phases of the Moon in his 1982 movie *E.T. the Extra-Terrestrial.* You have been selected to serve on a panel of scientists and attorneys to assist CSI in preparing their case. Your task is to gather relevant data, to synthesize your results into a strong argument for presentation to a court, and to propose potential solutions to the problem.

Read the following summary of the movie and take a look at the accompanying images from the scenes under investigation. Steven Spielberg uses the Moon effectively to create a mood in these scenes. He also uses the Moon to indicate the passing of time.

In a suburban community in California, a group of aliens are collecting plant specimens. When a pickup truck approaches, the aliens hastily leave. One alien who has become separated from the group is left behind. The men from the pickup truck pursue him, but he disappears into the darkness.

✪ Figure 8.5. The Boys Return to the Shed

That same evening, at around 10:00 p.m., a boy named Elliott hears a strange sound coming from the toolshed. A crescent Moon hanging low in the sky eerily illuminates the backyard. Frightened, Elliott runs into the house to enlist the support of his brother and his friends. They tease him when all they find at the shed are some footprints, which they mistakenly attribute to a coyote (see Figure 8.5).

✪ Figure 8.6. Elliott Goes Out by Himself at 2 a.m.

Later that night, at 2 a.m., while everyone else is sleeping, Elliott wanders off into a dry cornfield to search for the creature. The crescent Moon looms higher in the sky at this time (see Figure 8.6). Elliott follows footprints and the rustle of cornstalks. Suddenly he and the alien E.T. meet face to face, much to each other's surprise.

Your Challenge: *Is there evidence to support CSI's complaint about the inaccuracy of the movie? Explain in words and in a set of diagrams the accuracy of the movie's timeline of the phases of the Moon.*

PROBLEM 1

Page 2: More Information

E.T. the Extra-Terrestrial

It is four days later. Elliott, his older brother, and his younger sister have been hiding E.T. from their mother all this time. Meanwhile, E.T. has been busy building a machine to signal his spaceship to return and pick him up. He needs to go to an open place to use this device to "phone home." That evening, which happens to be Halloween, the boys sneak E.T. out of the house disguised as a ghost, pretending he is their little sister. After walking through their neighborhood of trick-or-treaters, Elliott takes E.T. into the forest in the front basket of his bike. E.T. takes control of the bike, driving it over a precipice, and boy, bike, and alien all become airborne as E.T. takes them flying over the forest and across the face of the rising full Moon (see Figure 8.7).

✪ Figure 8.7. E.T. Takes Elliott for a Ride Across the Face of the Moon

Your Challenge: *Is there evidence to support CSI's complaint about the inaccuracy of the movie? Explain in words and in a set of diagrams the accuracy of the movie's timeline of the phases of the Moon.*

PROBLEM 1

Page 3: Resources

E.T. the Extra-Terrestrial

Movie (Source Material for Problem)

- Spielberg, S. (Producer and Director). 1982. *E.T. The Extra-Terrestrial.* Universal City, CA: Universal Pictures.

Information About and Images of the Moon

- Photographs of the Moon's surface around the Apollo program landing sites, set up like Google Maps, are available at Google Moon: *www.google.com/Moon*.

- Animations of the Moon phases and the Moon orbiting Earth based on data from the Lunar Reconnaissance Orbiter are available at *http://astro.unl.edu/naap/lps/animations/lps.swf*.

- The U.S. Naval Observatory hosts the website Complete Sun and Moon Data for One Day at *http://aa.usno.navy.mil/data/docs/RS_OneDay.html*.

- StarDate, the public education and outreach arm of the University of Texas McDonald Observatory, has an online Moon Phase Calculator with daily pictures of the Moon as seen from Earth, plus a pictorial model, at *http://stardate.org/nightsky/Moon*.

- The Moon phase for any date and time can be found at the Moon Phase Images website: *http://tycho.usno.navy.mil/vphase.html*.

Materials for Modeling

- For the Sun, use a light source such as a bright bulb in a lamp without a shade, a flashlight, or an overhead projector.

- For the Moon, use a Styrofoam ball 4–6 cm in diameter (stuck on a bamboo skewer or pencil to act as a handle).

- For Earth, use your head.

- Optional: Use a sextant to determine the position of the Moon; see "Activity: Moon Finder" at *http://analyzer.depaul.edu/paperplate/Moon%20Finder.htm*.

PROBLEM 1

Teacher Guide

E.T. the Extra-Terrestrial

Problem Context

In the movie *E.T. the Extra-Terrestrial*, there are some scenes in which the boys are investigating noises in the shed and eventually discover E.T. In these scenes, which take place over a couple of weeks, the Moon is visible in the night sky. Was the director of the movie careful enough to make sure the phases shown match the time frame in which the story takes place? Students are asked by the Citizens for Scientific Integrity (a fictional organization created for the purpose of this problem) to do the fact-checking and write a letter to verify or refute the accuracy of the movie … or at least the Moon phases shown in the movie. See Table 8.1 for information about the timing of each scene.

Table 8.1. Moon Phase Data From the Movie

DATE	TIME	MOON PHASE AND POSITION
Day 1	~10 p.m.	Waning crescent, low in the sky
Day 2	2 a.m.	Waning crescent, higher in the sky
Day 5 or 6	~10 p.m.	Full Moon, low in the sky

Problem Solution

Model response:

Dear CSI members,

We find that you do have grounds for complaint about the lack of scientific accuracy in the way the Moon was depicted in the movie *E.T.* The half of the Moon facing the Sun reflects sunlight. Depending on the relative positions of the Sun, the Moon, and Earth, we see different amounts of the lit half of the Moon. The diagram below [Figure 8.8] shows the Moon orbiting Earth as seen from over the North Pole. Consider an alien standing on Earth. It will first be able to see the new Moon just as the Sun is rising (when it is at point A), and the new Moon will set as the Sun sets, because the Moon and the Sun are at approximately the same place in the sky. (In reality, the new Moon is difficult to see. The dark side is toward us and we have to look toward the Sun to see it.) The alien will first be able to see the full Moon as the Sun is setting, and full Moon will set as the Sun rises (when the alien is at point A) because the Sun and the Moon are in opposite directions. [SEP 2: Developing and Using Models; CC 2: Cause and Effect: Mechanism and Explanation]

Figure 8.8. Diagram of Moon Phases

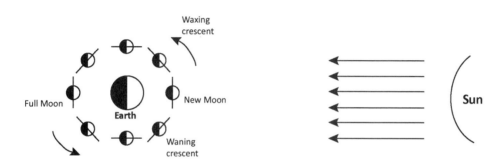

Lines through the Moon indicate the half facing Earth. (Earth and the Moon are not drawn to scale.)

The waning crescent shown in the early part of *E.T.* would really rise in the wee hours of the morning and set late in the afternoon. It would not be visible at 10 p.m. Since it takes 28 days for the Moon to orbit Earth and therefore to cycle through its phases, each of the pictures of the Moon in the diagram represents how far the Moon moves in $3\frac{1}{2}$ days. Therefore, a full Moon appears roughly 15–17 days after a thin waning crescent, not 5 days. See the diagram. [Figure 8.8] [SEP 5: Analyzing and Interpreting Data]

Activity Guide

The simplest way for students to make sense of this problem is to develop a model. This is most easily done in a darkened room with a bright, bare bulb simulating the Sun. The Sun can also be represented by an overhead projector or flashlights. However, students need to be reminded that, unlike the Sun, these sources of light are directional. Whatever the source of illumination, students take on Earth's perspective (their heads represent Earth). A small foam ball on a skewer or pencil can represent the Moon. (The skewer or pencil allows students to hold the "Moon" without their hands getting in the way of the demonstration.)

Students can simulate the Moon (Styrofoam ball) revolving counterclockwise around Earth (their head) and observe patterns in the amount of the illuminated part of the ball they see. They need to record how much they see along with the relative positions of the Sun and Earth. One way to record these observations is to shade in circles to show the phase of the Moon at each position. Note that at full Moon, students' heads will cast a shadow on their Moon (an eclipse), unless they hold their Moon up fairly high. (Eclipses do not occur every full Moon, because the orbital planes of the Moon and Earth are not aligned. For more on this topic, see Problem 3.)

This activity can be extended by having students observe the actual Moon. These observations will help students overcome misconceptions such as the common belief that the Moon, like the stars, is only visible at night. (Because the Moon is so much closer to us than the stars, it is bright enough to be seen any time of day when an observer's side of Earth is facing it.) Students can make semiquantitative observations of the Moon using a sextant to measure how high in the sky it is and how far east and west of south it is. (See "Materials for Modeling" on Page 3: Resources.) Students should be able to explain their observations using their models.

SAFETY PRECAUTIONS

1. Safety glasses or safety goggles are required for this activity.

2. Lamps produce heat, which can burn skin. Handle with care.

3. Keep lamps away from any water source to prevent shock.

4. Skewers are sharps and can puncture or cut skin. Handle with care.

5. Make sure any fragile items or trip/fall hazards are removed from the work area.

PROBLEM 1

Assessment

E.T. the Extra-Terrestrial

Transfer Task

On July 20, 1969, Apollo 11 became the first manned mission to land on the Moon and Neil Armstrong became the first human to step on another planetary body. From the Moon, Armstrong could clearly see Earth. Reflecting on this experience, Armstrong later said that, "It suddenly struck me that that tiny pea, pretty and blue, was Earth. I put up my thumb and shut one eye, and my thumb blotted out the planet Earth. I didn't feel like a giant. I felt very, very small" (*Clovis News Journal* 2009). What did Neil Armstrong see when he looked at Earth? Specifically, what was the phase of Earth?

You may use textbooks to research your answer, but the only internet resource you may use is the Moon Phase Calculator (*http://stardate.org/nightsky/Moon*). For a complete analysis, please remember to do all of the following: state the phase of Earth in words (including waxing/waning or first/third, as appropriate); draw Earth as it would appear from the Moon (draw the illuminated portion only); explain how you deduced this phase; and sketch a well-labeled diagram that supports your answer.

> *Model response:* On July 11, 1969, the Moon was several days past new. This means that the Moon, the Sun, and Earth were almost in a line as seen from somewhere over the North Pole. At new Moon, people on the Moon would see all of the illuminated side of Earth. A few days later, they would still see most of the illuminated side of Earth, but the right-hand side would not be in the Sun's light. Earth would have appeared to be in its waning gibbous phase. The diagram below [Figure 8.9] shows the phases and relative positions of the Sun, the Moon, and Earth on that day. [CC 1: Patterns]

Figure 8.9. Diagram of Moon Phase on the Day Neil Armstrong Walked on the Moon

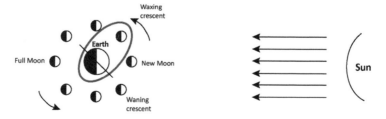

Line through Earth indicates the half facing the Moon. (Earth and the Moon are not drawn to scale.)

Application Question

One pleasant spring evening from their backyard in Michigan, Brent and his younger brother, Ethan, noticed the Moon shining high in the sky (see photo [Figure 8.10]). Ethan looked at it thoughtfully and then asked his brother the following questions. Please help Brent by answering them and explaining why.

⊗ **Figure 8.10. Waxing Crescent Moon**

a. "Will I see the same Moon tomorrow morning?"

b. "What Moon will people in Australia see tonight?"

c. "My favorite is the full Moon. How long do I have to wait to see the full Moon?"

Model response:

a. Because Earth spins on its axis once each day, celestial objects, including the Moon, appear to rise in the east, move across the sky, and set in the west. For objects that pass directly overhead, this takes about 12 hours. The waxing crescent Moon that Brent and Ethan are looking at rose before noon and will set before midnight; therefore, it won't be visible in the morning. [CC 1: Patterns]

b. Because the phase of the Moon depends on the relative positions of the Sun, the Moon, and Earth, everyone on Earth sees the same phase of the Moon on any given day or night. [CC 2: Cause and Effect: Mechanism and Explanation]

c. Brent and Ethan are looking at a waxing crescent Moon that occurs between a new Moon and a full Moon. This means that they have to wait more than a quarter of the Moon's cycle to see the full Moon. Because it takes the Moon about 28 days to orbit Earth, this means they must wait more than 7 days before they see the full Moon. [CC 1: Patterns]

Reference

Clovis News Journal. 2009. Armstrong Last of Heroes With Integrity. July 20.

Problem 2: Obsidian Sun

Alignment With the *NGSS*

PERFORMANCE EXPECTATIONS	• *MS-ESS1-1:* Develop and use a model of the Earth-Sun-Moon system to describe the cyclic patterns of lunar phases, eclipses of the Sun and Moon, and seasons.
SCIENCE AND ENGINEERING PRACTICES	• Developing and Using Models • Analyzing and Interpreting Data • Using Mathematics and Computational Thinking • Constructing Explanations and Designing Solutions • Obtaining, Evaluating, and Communicating Information

DISCIPLINARY CORE IDEAS	• *ESS1.A: The Universe and Its Stars*
	    ○ *Grades 6–8:* Patterns of the apparent motion of the Sun, the Moon, and stars in the sky can be observed, described, predicted, and explained with models.
	• *ESS1.B: Earth and the Solar System*
	    ○ *Grades 6–8:* The solar system consists of the Sun and a collection of objects, including planets, their Moons, and asteroids that are held in orbit around the Sun by its gravitational pull on them. This model of the solar system can explain tides, eclipses of the Sun and the Moon, and the motion of the planets in the sky relative to the stars.

CROSSCUTTING CONCEPTS	• Patterns • Scale, Proportion, and Quantity • Systems and System Models

Keywords and Concepts

Solar eclipse

Problem Overview

Students try to explain what happens in a solar eclipse, and why a person in one location can see a solar eclipse while a person in another place cannot.

Images available in full color on the Extras page (*www.nsta.org/pbl-earth-space*) are marked with the following icon: ✪.

PROBLEM 2

Page 1: The Story

Obsidian Sun

Lakesha was annoyed. It was a beautiful, sunny Saturday morning and she had a bad cold. Her mother, who was out running errands, had left her with some hot tea. Lakesha had moved to Oregon the previous summer and was finally starting to feel attached to her new home and school. As she settled down on the couch, she wondered what her friends were doing. She started checking social media with the TV playing in the background.

Just as she started to pull up some social media pages from friends from her old school, Lakesha heard the TV announcer say that a solar eclipse had begun, and it would be total in 30 minutes. She hadn't noticed a thing: The Sun was shining brightly through the window, making a long shadow of her cup across the table. She wanted to watch the eclipse progress, but she knew she wasn't supposed to look directly at the Sun. The TV announcer said to use special glasses, but Lakesha only had regular sunglasses. Fortunately, the announcer demonstrated another way to observe the eclipse. Lakesha quickly followed his example. She was determined that the day wouldn't be a complete waste. Using her mother's darning needle, she poked a hole in a piece of cardboard. She held the cardboard up to the window where the sunlight was streaming in and could see an image of the Sun on the floor. She put a piece of blank paper down on the floor where the image was to act like a screen. The image was small, but clear.

Lakesha was delighted! She watched the dark Moon slowly scroll across the face of the Sun. She wondered if her mother noticed what was happening. She thought of texting her mother, but decided against it since her mother was likely to be driving. On a whim, she called her friend Miranda from her old neighborhood in Florida.

Miranda answered right away. "Lakesha, how are you, girl? Is anything wrong? I haven't talked to you in so long. …"

"I'm fine! Well, no, I have a bad cold. But that's not why I called. Listen. Look out your window! There's an eclipse of the Sun!"

"Well, it's a nice sunny day down here," Miranda said. "I was just outside."

Lakesha was confused. Why wasn't Miranda seeing what she saw? She quickly explained how to view the eclipse safely. "Now go look, quick!"

A few moments later, when Miranda called back, she said: "Hey, you're right! There is an eclipse! I only see it when I project the image, as you told me. Otherwise I wouldn't even notice. But it's not total. The Sun looks like a cookie with just one tiny bite taken out of it. But hey, while I got you on the line, you'll never guess what happened last week. ..."

As Lakesha listened to Miranda, she noticed that the Moon completely covered the Sun. Birds sang their evening songs. It appeared that night had fallen. Stars came out. The obsidian Sun was fringed with dark streamers.

After less than three minutes of totality, a bit of the Sun peeked around the edge of the Moon, shining like a diamond. Lakesha interrupted Miranda to ask what she was seeing. Miranda said things looked pretty much the same as before.

They chatted another half hour or so, until Lakesha's eclipse was completely over. At that time, Lakesha again asked Miranda what she could see. Miranda reported that the Moon took a bigger bite out of the Sun—not much more than half the Sun, but that now was getting smaller again. Lakesha was confused.

Your Challenge: *Use a model to explain to Lakesha why she saw a total solar eclipse but Miranda didn't.*

PROBLEM 2

Page 2: More Information

Obsidian Sun

Lakesha lives in Happy Valley—a suburb of Portland, Oregon. She moved there from Orlando, Florida, where her friend Miranda still lives. Portland and Orlando are about 3,000 miles (4,828 km) and three time zones apart.

Thirty minutes after Lakesha witnessed totality in Portland, Miranda observed a maximum partial eclipse, with only 50% of the Sun covered by the Moon at 11:48 a.m. Eastern Standard Time (EST).

The events in this story are based on the total solar eclipse that took place on February 26, 1979. Totality occurred in Portland between 8:13 and 8:15 a.m. Pacific Standard Time (PST). In Orlando, the eclipse was only a partial eclipse.

Your Challenge: *Use a model to explain to Lakesha why she saw a total solar eclipse but Miranda didn't.*

PROBLEM 2

Page 3: Resources and Investigation

Obsidian Sun

Resources

- Students can view data and maps related to the total solar eclipse of February 26, 1979 at *http://eclipse.gsfc.nasa.gov/SEgoogle/SEgoogle1951/SE1979Feb26Tgoogle.html*.

- Students can search for data about eclipses in any year at the HMNAO (Her Majesty's Nautical Almanac Office) Eclipse Portal: *www.eclipse.org.uk/eclbin/query_hmnao.cgi*.

- NASA's website offers resources, images, and videos about eclipses: *www.nasa.gov/topics/solarsystem/features/eclipse/index.html*.

- NASA also provides information on how to safely view a solar eclipse: *http://eclipse.gsfc.nasa.gov/SEhelp/safety2.html* or at *www.exploratorium.edu/eclipse/how-to-view-eclipse*.

Solar Eclipse Model Investigation

MATERIALS (PER GROUP)

- For the Sun, use a light source such as a flashlight or a lamp without a shade.

- For the Moon, use a Styrofoam ball 4–6 cm in diameter (stuck on a bamboo skewer or pencil to act as a handle).

- For Earth, use your head.

PROCEDURE

1. Experiment with the position of Earth (your head) and the Moon (Styrofoam ball) to see what relative distance the Moon would have to be from the Sun (light source) and your location on Earth (your eyes) in order to completely "eclipse" the Sun. Measure the distances between Earth, the Moon, and the Sun. Have different group members experiment with being Earth. Do you all get roughly the same distances?

2. To view the size of the shadow cast by the Moon on Earth, place a white piece of paper where your head was. Draw around the edge of the shadow and measure its diameter.

3. Make a scale drawing of Earth, the Sun, and the Moon positioned during a total solar eclipse (as seen from somewhere above the North Pole). Pick a scale that will work both for the distances and the size of the Moon and its shadow.

PROBLEM 2

Teacher Guide

Obsidian Sun

Problem Context

In this problem, two friends are talking about solar eclipses. They notice that only one of them can see a solar eclipse on a given day because they live far from each other. To help them understand why they don't both see the same eclipse, students should investigate what causes a solar eclipse, how often solar eclipses occur, and the size of the shadow cast on Earth's surface.

Problem Solution

> *Model response:* A solar eclipse occurs when there is a new Moon; that is, when the Moon moves between Earth and the Sun, blocking the view of the Sun at least partially. [CC 2: Cause and Effect: Mechanism and Explantion] If the plane of the Moon's orbit around Earth and the plane of Earth's orbit around the Sun were the same, we would have a total eclipse every month. However, the Moon's orbital plane is tilted relative to Earth's, so most months the shadow of the Moon does not fall on Earth. When it does fall on Earth, it is too small to cover all of Earth. People like Lakesha who are in the path of the shadow, see a total eclipse. People like Miranda who are close to, but not in, the path of the shadow see a partial eclipse. Other people will see no eclipse at that time. [CC 1: Patterns]

Activity Guide

Students can build a model of the Earth-Moon-Sun system using Styrofoam balls and a light source, as described in the Page 3: Resources section of this problem. Modeling the Earth-Moon-Sun system works best in a darkened room. If you are using lamps without shades to represent the Sun, students may be confused by light and shadows from lamps other than their own. It may be better to use one light in the center of the room that is everyone's "Sun." The students will create a total eclipse when the Moon (ball) is between themselves and the Sun (light). Due to the small size of the ball representing the Moon (the ratio between the Styrofoam balls and the size of a child's head is roughly the same as the ratio of the diameter of the Moon to the diameter of Earth), students will need to be fairly far back from the Sun to see a total eclipse.

As an alternative to building a model, have students draw a picture of the arrangement of Earth, the Moon, and the Sun at the time of a solar eclipse without worrying about scale.

Viewing a solar eclipse can also supplement the experiences in this book, but they are hard to include in lesson plans because they depend on clear skies, timing with an eclipse event, and proper equipment or preparation. If school is in session during a solar eclipse (see *www.timeanddate.com/eclipse/list.html* for a schedule of eclipses over the next 10 years), you could have your students view an actual eclipse. Even if there will not be an eclipse when school is in session, you can teach them how to view an eclipse safely, for future use.

Warning: The intense visible light from the Sun can damage the retina of the eye if the Sun is viewed directly. This can cause blind spots on the retina even if you are not feeling discomfort. It is *only* advisable to view the eclipse in one of three safe ways. The first way is to shield the eyes with welding goggles (rated 14 or higher) or special eclipse glasses. Some telescopes also include a solar filter that lets viewers look at a solar eclipse. Another option is to build a pinhole projector to view the eclipse indirectly. This pinhole projector can be used to safely view the Sun at any time. The pinhole projector consists of a sheet of cardboard with a small hole in it and a screen (something white and smooth such as a white poster board) that can be positioned about a yard behind the pinhole. Students can get instructions for making a pinhole projector at this website: *www.exploratorium.edu/eclipse/how-to-view-eclipse*.

Another device that can allow students to view an eclipse is a Sunspotter, available from Scientifics Direct (*www.scientificsonline.com/product/Sunspotter*). This device is designed for viewing sunspots, dark areas of activity on the Sun's surface. But it also lets students see an eclipse.

Nature creates its own pinholes such as the gaps between leaves of a tree. Figure 8.11 (p. 208) shows the ground under a tree during an almost total eclipse. A pinhole projector, a Sunspotter, or a tree with leaves offers a way for students to view the Sun even when no solar eclipse is occurring.

SAFETY PRECAUTIONS

1. Safety glasses or safety goggles are required for this activity.

2. Skewers and pushpins are sharps and can puncture or cut skin. Handle with care.

3. Lamps produce heat, which can burn skin. Handle with care.

4. Keep lamps away from any water source to prevent shock.

5. Make sure any fragile items or trip/fall hazards are removed from the work area.

✪ Figure 8.11. Solar Eclipse Seen Through a Filter of Leaves

Note: A full-color version of this image is available on the Extras page (*www.nsta.org/pbl-earth-space*).

NATIONAL SCIENCE TEACHERS ASSOCIATION

PROBLEM 2

Assessment

Obsidian Sun

Transfer Task

Lakesha and Miranda wonder if it is possible for other planets to cause an eclipse visible from Earth. Based on your understanding of the cause of a solar eclipse, what would the girls expect to see if another planet passes between the Sun and Earth? Make sure to include what planets might be able to do this and how we could view the event.

Model response: There are only two planets that pass between Earth and the Sun: Mercury and Venus. The other planets move in an orbit that takes them farther from the Sun than Earth, so they never pass "in front of" Earth. [CC 1: Patterns; CC 4: Systems and System Models]

But neither of these planets can cause an eclipse. Although both Mercury and Venus are larger than the Moon, they are so far away from Earth that we would not notice a change in the light levels. These events are called transits. [SEP 6: Constructing Explanations and Designing Solutions; CC 3: Scale, Proportion, and Quantity; for a photograph and explanation of transits, go to *www.nasa.gov/mission_pages/sunearth/news/2012-venus-transit.html#.V5JFV3puNpU*]

Problem 3: Copper Moon

Alignment With the *NGSS*

PERFORMANCE EXPECTATIONS	• *MS-ESS1-1:* Develop and use a model of the Earth-Sun-Moon system to describe the cyclic patterns of lunar phases, eclipses of the Sun and Moon, and seasons.
SCIENCE AND ENGINEERING PRACTICES	• Developing and Using Models • Analyzing and Interpreting Data • Using Mathematics and Computational Thinking • Constructing Explanations and Designing Solutions • Obtaining, Evaluating, and Communicating Information

DISCIPLINARY CORE IDEAS	• *ESS1.A: The Universe and Its Stars*
	  ○ *Grades 6–8:* Patterns of the apparent motion of the Sun, the Moon, and stars in the sky can be observed, described, predicted, and explained with models.
	• *ESS1.B: Earth and the Solar System*
	  ○ *Grades 6–8:* The solar system consists of the Sun and a collection of objects, including planets, their Moons, and asteroids that are held in orbit around the Sun by its gravitational pull on them. This model of the solar system can explain tides, eclipses of the Sun and the Moon, and the motion of the planets in the sky relative to the stars.

CROSSCUTTING CONCEPTS	• Patterns • Scale, Proportion, and Quantity • Systems and System Models

Keywords and Concepts

Lunar eclipse

Problem Overview

Two friends living across the country from each other see a lunar eclipse at the same time and wonder how it is possible.

Images available in full color on the Extras page (*www.nsta.org/pbl-earth-space*) are marked with the following icon: ☉.

PROBLEM 3

Page 1: The Story

Copper Moon

Lakesha had moved from Florida to Oregon the previous summer but still missed her friends in Florida. One night her phone rang while she was sleeping. She grabbed it, wondering who would be calling at 3:30 a.m. It turned out to be Miranda, one of her friends from Florida.

"You're never going to believe this!" Miranda screamed from the other end of the line. "Look outside your window!"

"Miranda," Lakesha cried, "it must be 6:30 a.m. for you. What's going on?"

"Everything's OK. I've been up all night working on my robot for class. But I know you are so interested in eclipses, so I thought you'd like to see another one! I've been watching this one for an hour. I'm looking at the Moon right now. It's totally eclipsed. It's gorgeous! Look out the window!"

Lakesha muffled a long yawn. "What am I going to do with you, girl?" Then she laughed. "OK. Let me take a look." She padded barefoot across her room to her window. And there, high in the southwest, she saw a copper Moon in a velvet sky. They watched together for a few minutes, speaking softly, until Miranda had to go to sleep. But Lakesha leisurely enjoyed the rest of the show.

Your Challenge: *Use a model to explain why it was possible that Lakesha and Miranda could leisurely watch the same lunar eclipse together, but not a solar eclipse.*

PROBLEM 3

Page 2: More Information

Copper Moon

On September 6, 1979, there was a total lunar eclipse. Observers would have been able to see the lunar eclipse without special equipment, as long as the sky was clear in their location.

It was visible to most observers in the Western Hemisphere. The eclipse began at 5:18 a.m. Eastern Daylight Time (EDT) when the Moon moved into Earth's shadow. Many observers report that the Moon looks bigger or "more three-dimensional" once an eclipse begins. On September 6, 1979, the Moon was totally eclipsed when it moved through the darkest part of Earth's shadow, which occurred between 6:32 and 7:18 a.m. EDT. The eclipse was completely over when the Moon exited Earth's shadow at 8:30 a.m. EDT. The later part of the eclipse was not visible to Miranda in Florida.

Your Challenge: *Use a model to explain why it was possible that Lakesha and Miranda could leisurely watch the same lunar eclipse together, but not a solar eclipse.*

PROBLEM 3

Page 3: Resources and Investigation

Copper Moon

Resources

- Students can search for data about eclipses in any year at the HMNAO (Her Majesty's Nautical Almanac Office) Eclipse Portal: *www.eclipse.org.uk/eclbin/query_hmnao.cgi*.

- NASA's website offers resources, images, and videos about eclipses: *www.nasa.gov/topics/solarsystem/features/eclipse/index.html*.

- Information about the total lunar eclipse on September 6, 1979, is available at *www.timeanddate.com/eclipse/lunar/1979-september-6*.

Lunar Eclipse Model Investigation

MATERIALS (PER GROUP)

- For the Sun, use a light source such as a flashlight or a lamp without a shade.

- For the Moon, use a Styrofoam ball 4–6 cm in diameter (stuck on a bamboo skewer or pencil to act as a handle).

- For Earth, use your head.

PROCEDURE

1. Experiment with relative positions of the Sun (light), the Moon (Styrofoam ball), and Earth (your head) until you see a lunar eclipse.

2. Draw a diagram of the arrangement as seen from somewhere above the North Pole.

3. How does this arrangement compare with that of a solar eclipse?

PROBLEM 3

Teacher Guide

Copper Moon

Problem Context

In this problem, two friends are both watching a lunar eclipse while talking on the phone, even though they live on opposite sides of the country. Solar eclipses are only visible across a small area, so they wonder why they are both able to see the same lunar eclipse at the exact same moment. To help them understand this comparison, students can investigate a lunar eclipse and how often these eclipses occur. The model-building activity described in the "Activity Guide" section can be modified to demonstrate both solar and lunar eclipses.

Problem Solution

Model response: Lunar eclipses occur at the time of the full Moon when the Moon moves into Earth's shadow. As with solar eclipses, lunar eclipses occur only in months when the full Moon is in the same plane as Earth and the Sun. Because the shadow of Earth is big compared to the Moon, all of the Moon is in shadow for a significant amount of time. [CC 2: Cause and Effect: Mechanism and Explanation]

Any people on Earth who can see the Moon when it is Earth's shadow will see the same event. Some may see the Moon high in the sky, others closer to the horizon, but the eclipse will be visible from the entire "night" side of Earth. This is why the two girls see the same eclipse at the same time. [CC 1: Patterns]

Activity Guide

Students can simulate a lunar eclipse using the lamp-ball-head model described on Page 3: Resources. When they place their head between the light and the ball and hold the ball at the right height, their shadow will cover the ball (the Moon). Both eyes (each representing a different location) will see the ball in shadow.

Students may also have an opportunity to see a lunar eclipse during the school year. No special equipment is needed to view a lunar eclipse.

SAFETY PRECAUTIONS

1. Safety glasses or safety goggles are required for this activity.

2. Lamps produce heat, which can burn skin. Handle with care.

3. Keep lamps away from any water source to prevent shock.

4. Skewers are sharps and can puncture or cut skin. Handle with care.

5. Make sure any fragile items or trip/fall hazards are removed from the work area.

PROBLEM 3

Assessment

Copper Moon

Transfer Task Several months after the eclipse Lakesha and Miranda had seen, they read on the internet that there would be another eclipse, so they decided to stay up and watch while talking on the phone again. The eclipse they saw looked very different this time. At the moment when the largest part of the Moon was in Earth's shadow, it looked like the image shown in Figure 8.12.

✪ **Figure 8.12. Partial Lunar Eclipse**

Explain why the entire Moon is not darkened in the eclipse. Include in your explanation the position of the Moon relative to Earth and the Sun.

Model response: The image shown is a partial lunar eclipse. In this eclipse, only part of the Moon passes into the shadow of Earth. Because the lower part of the Moon is dark, this image shows that the plane of the Moon's orbit at the time the girls saw the eclipse was not exactly in line with the path of light from the Sun to Earth. In most months, the Moon does not pass into the shadow at all, so the entire face of the Moon facing Earth is illuminated. [SEP 6: Constructing Explanations and Designing Solutions; CC 1: Patterns; CC 2: Cause and Effect: Mechanism and Explanation]

Application Question

Daniella watched a total lunar eclipse at her home near Pittsburgh, Pennsylvania, on the night of February 20, 2007. Totality began at 10:01 p.m. At that time, Daniella called her cousin Sheri, in New Orleans, Louisiana, and told her to go outside and look at the eclipse.

a. What, if anything, was Sheri able to see at that time? Explain.

b. A half hour later, Daniella decided to call her Aunt Willa in New York. What, if anything, was Aunt Willa able to see at that time? Explain.

Model response:

a. Lunar eclipses occur when the Moon moves into the shadow of Earth. Earth's shadow is big enough to cover the entire Moon. Just as everyone on Earth sees the same phase of the Moon, everyone on Earth sees the same eclipse if they are on the part of Earth facing the Moon. [CC 1: Patterns] (The exception to this is if the eclipse occurs before the Moon rises or after it sets for some people.)

b. Aunt Willa is farther east, so the Moon will be lower in the sky, but she will see the same stage of the eclipse as Daniella. Because totality lasts an hour or more, the eclipse will probably be total. [CC 1: Patterns]

Problem 4: Morning Star, Evening Star

Alignment With the *NGSS*

PERFORMANCE EXPECTATIONS	• *MS-ESS1-1:* Develop and use a model of the Earth-Sun-Moon system to describe the cyclic patterns of lunar phases, eclipses of the Sun and Moon, and seasons.
SCIENCE AND ENGINEERING PRACTICES	• Developing and Using Models • Analyzing and Interpreting Data • Using Mathematics and Computational Thinking • Constructing Explanations and Designing Solutions • Obtaining, Evaluating, and Communicating Information

DISCIPLINARY CORE IDEAS	• *ESS1.A: The Universe and Its Stars*
	  ○ *Grades 6–8:* Patterns of the apparent motion of the Sun, the Moon, and stars in the sky can be observed, described, predicted, and explained with models.
	• *ESS1.B: Earth and the Solar System*
	  ○ *Grades 6–8:* The solar system consists of the Sun and a collection of objects, including planets, their Moons, and asteroids that are held in orbit around the Sun by its gravitational pull on them. This model of the solar system can explain tides, eclipses of the Sun and the Moon, and the motion of the planets in the sky relative to the stars.

CROSSCUTTING CONCEPTS	• Patterns • Systems and System Models • Stability and Change

Keywords and Concepts

Movement of planets relative to Earth

Problem Overview

A brother and sister learn that Venus is visible as either a morning star or an evening star at different times of the year, and they try to explain the pattern of movement of Venus in the sky.

Images available in full color on the Extras page (*www.nsta.org/pbl-earth-space*) are marked with the following icon: ✪.

PROBLEM 4

Page 1: The Story

Morning Star, Evening Star

Isaiah and his sister, Tamila, were awestruck by the stone pyramid they were visiting in Mexico. As they gazed at the steep steps rising out of the jungle, their guide told them about the traditional ball game that the ancient Mayans played. The winner was sacrificed to the gods. Isaiah found it hard to believe that anyone would try to win under such circumstances, but the guide explained that it was considered a heavenly honor.

Isaiah and Tamila trailed after their group as the guide started to explain about the Mayan calendar. "We developed an accurate calendar based on the appearance of Venus in the morning and evening skies. Our ancestors understood the patterns of the stars very well."

The guide went on to explain that sometimes Venus outshines every star in the evening sky. For several months, on each successive evening, Venus climbs higher and gets brighter in the west after sunset. Then it descends and gets dimmer for several months before appearing to dive right into the Sun, disappearing for about a week. It reappears in the morning, climbing higher in the east before sunrise, getting brighter each morning, and then gradually sinking and dimming as it dives again into the rising Sun. Then after two months, it reappears as an evening star.

"This cycle of Venus was well known to our ancestors," the guide continued, "and became the basis for a 584-day calendar. Do you see this figure here above the temple door?" He pointed to a stucco relief of a man wearing an eagle helmet with legs extended backward, as if he were about to jump into a pool of water. "This figure is common here at Tulum as well as other Mayan sites. We call him the 'Diving God,' and many of us believe he represents Venus at that moment it disappears into the Sun."

This surprised Isaiah and Tamila. In school they had learned that Venus orbits the Sun in a lot less than a year. Why did the guide say Venus has a 584-day cycle? And how come it is only seen in the morning and evening?

Your Challenge: *Use words and a picture to explain to Isaiah and Tamila why Venus appears only in the morning and evening and why it has a 584-day cycle.*

PROBLEM 4

Page 2: More Information

Morning Star, Evening Star

On the following pages are eight sets of observations made over eight consecutive months, from January 2007 through August 2007 (see Figures 8.13–8.20, pp. 220–222). Each image on the left is a computer-generated picture of the western horizon just after sunset. Each image on the right is a close-up picture of Venus taken just after sunset as it would appear through a simple telescope (except for Figure 8.20 [p. 222]—there is no picture of Venus for August 2007 because the planet was not visible then). *Note:* All of these images are screen captures using Stellarium (*www.stellarium.org*), a free open-source planetarium.

Your Challenge: *Use words and a picture to explain to Isaiah and Tamila why Venus appears only in the morning and evening and why it has a 584-day cycle.*

✪ Figure 8.13. Sunset on Earth Looking Due West (left) and Close-Up Picture of Venus (right), January 2007

✪ Figure 8.14. Sunset on Earth Looking Due West (left) and Close-Up Picture of Venus (right), February 2007

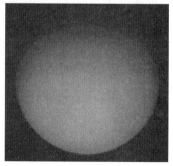

✪ **Figure 8.15. Sunset on Earth Looking Due West (left) and Close-Up Picture of Venus (right), March 2007**

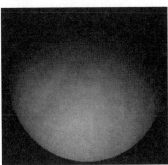

✪ **Figure 8.16. Sunset on Earth Looking Due West (left) and Close-Up Picture of Venus (right), April 2007**

✪ **Figure 8.17. Sunset on Earth Looking Due West (left) and Close-Up Picture of Venus (right), May 2007**

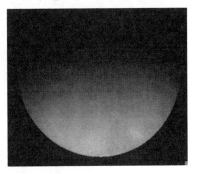

✪ Figure 8.18. Sunset on Earth Looking Due West (left) and Close-Up Picture of Venus (right), June 2007

✪ Figure 8.19. Sunset on Earth Looking Due West (left) and Close-Up Picture of Venus (right), July 2007

 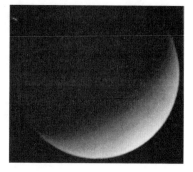

✪ Figure 8.20. Sunset on Earth Looking Due West, August 2007

PROBLEM 4

Page 3: Resources

Morning Star, Evening Star

- The interactive Sky Map at Your Sky, an interactive planetarium on the web, lets students find the location of many stars and planets on any day: *www.fourmilab.ch/yoursky*.

- Stellarium is a free open-source planetarium that shows images of objects in space for any date: *www.stellarium.org*.

- NASA's website has information, images, and videos of Venus: *http://solarsystem.nasa.gov/planets/venus*.

- Nine Planets is another website with information about Venus: *www.nineplanets.org/venus.html*.

PROBLEM 4

Teacher Guide

Morning Star, Evening Star

Problem Context

The Mayans are one of many ancient civilizations that based their calendar on the motion of the stars and planets. Their architecture was designed to mark important astronomical observations, and their priests consulted the stars about when to plant crops. Isaiah and Tamila are confused about the 584-day cycle of Venus when its orbit is less than a year. They are curious about the Mayans' observations, and they want to know more.

Problem Solution

Model response: Venus's orbit is closer to the Sun than ours, so no matter where it is in its orbit, Venus is always in the direction of the Sun from us. This means that it is often above the horizon during the day when the planet is not bright enough to be seen in the sunlit sky. For small parts of its orbit, Venus is visible from Earth during morning or evening twilight. See the diagram below. [Figure 8.21]

Advanced response: The amount of time it takes before Earth, Venus, and the Sun are in the same relative positions is 584 days. In the diagram below [Figure 8.21], Venus is positioned relative to Earth so that it is visible in the evening around sunset. In 225 days (the length of a Venusian year), Venus would be back in the same place. However, Earth would be 225/365 (about 0.6) of the way around its orbit. At day 365, Earth would be back in the same place, but Venus would be (365 − 225)/225 or about 0.6 of the way around its second orbit. At day 584, Earth would make 584/365 or 1.6 orbits. Venus would make 584/225 or

Figure 8.21. Orbits of Earth and Venus

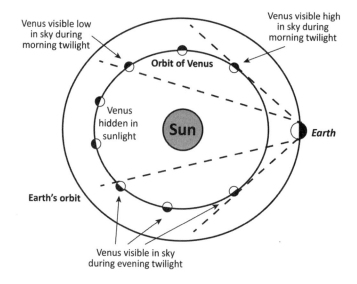

2.6 orbits. Both Earth and Venus would be 0.6 of the way around the orbit from the original position so that their positions relative to the Sun would be the same. Rather than working this out mathematically, it is easier to see using a model and considering the relative positions of Earth, Venus, and the Sun after a particular amount of time. [SEP 5: Using Mathematics and Computational Thinking; SEP 7: Engaging in Argument From Evidence; CC 1: Patterns; CC 2: Cause and Effect: Mechanism and Explanation; CC 4: Systems and System Models]

PROBLEM 4

Assessment

Morning Star, Evening Star

Transfer Tasks

TRANSFER TASK 1

Sometimes Mars is visible throughout the night for many months. In December 2007, Mars could be seen high in the sky at midnight. Explain how Mars can be visible in the middle of the night but Venus cannot, and support your explanation with a well-labeled diagram.

Model response: Mars's orbit around the Sun is bigger than Earth's. When Mars and Earth happen to be near each other in their orbits, Mars will be on the side of Earth away from the Sun. This means we will see it at night. Venus's orbit is closer to the Sun than ours, so no matter where it is in its orbit, Venus is always in the direction of the Sun from us. This means that it is often above the horizon during the day when it is not bright enough to be seen in the sunlit sky. For small parts of its orbit, Venus is visible from Earth during morning or evening twilight. See the diagram. [Figure 8.22] [CC 1: Patterns; CC 3: Scale, Proportion, and Quantity; CC 4: Systems and System Models]

Figure 8.22. Orbits of Earth, Venus, and Mars

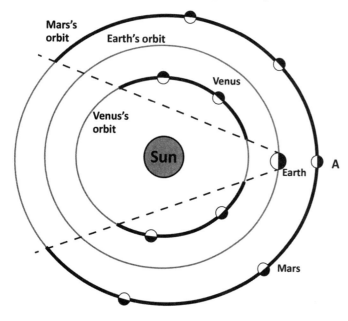

TRANSFER TASK 2

Discuss the cycle of phases of Mars as observed from Earth through a telescope. Accompany your explanation with a well-labeled diagram showing Mars in at least four different, equally spaced positions in its orbit. To simplify the analysis, assume a stationary Earth.

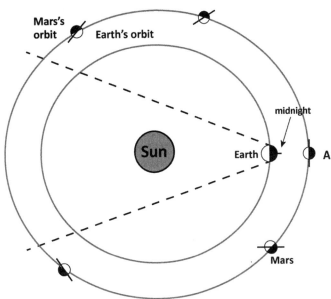

Figure 8.23. Phases of Mars

Model response: Like the Moon, all of the planets are visible because they reflect sunlight. Their phase—that is, how much of the face toward us we see—depends on the relative positions of the Sun, Earth, and planet.

When Mars and Earth happen to be near each other in their orbits, Mars will be on the side of Earth away from the Sun and will be visible at night. The side of Mars facing Earth will be fully lit so that Mars (as seen through a telescope) will appear to be full. When Mars and Earth are opposite each other in their orbits, Mars is on the far side of the Sun and not visible from Earth. Earth faces it only during the day when it is not visible in the lit sky. When Mars is a quarter of the way around its orbit relative to Earth, it will be visible from Earth. More than half of the face of Mars toward us will be lit and visible. See the diagram. [Figure 8.23] [SEP 6: Constructing Explanations and Designing Solutions; CC 1: Patterns]

Application Question

One evening after sunset, Crystal noticed Venus shining brightly high in the west. Sketch what Crystal saw when she looked at Venus through binoculars. [Figure 8.24] Explain your drawing.

Figure 8.24. Girl in Thought

Model response: With binoculars, we can see the phase of Venus. If Crystal is looking west at sunset, the Sun is just below the horizon in the direction in which she is looking. This means that the lower half of Venus (the half toward the Sun) is illuminated. See the drawing below. [Figure 8.25] [CC 1: Patterns]

Figure 8.25. Drawing of Phase of Venus

Astronomy Problems: General Assessment

General Questions

General Question 1

Why do we see the Moon go through phases, and why do the phases change? Explain clearly in words and support your explanation with a well-labeled diagram.

> *Model response:* The Moon is visible because it reflects light from the Sun. Observers on Earth only see the part of the Moon that is illuminated by the Sun and that is facing Earth. [CC 2: Cause and Effect: Mechanism and Explanation] The phase of the Moon (the portion of the Moon seen from Earth) changes as the Moon orbits Earth. [CC 1: Patterns] See the diagram below. [Figure 8.26]

Figure 8.26. Diagram of Moon Phases

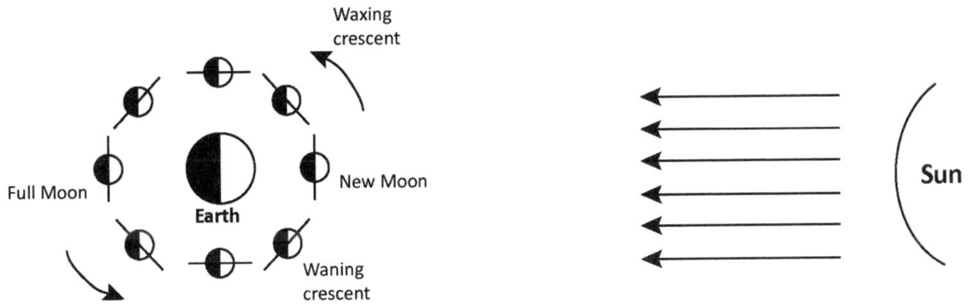

Lines through the Moon indicate the half facing Earth. (Earth and the Moon are not drawn to scale.)

General Question 2

Why is Venus only seen in the morning or in the evening, but never in the middle of the night? Explain clearly in words and support your explanation with a well-labeled diagram.

> *Model response:* The location of a planet in the sky is determined by a planet's position in its orbit relative to Earth at any specific time. For example, Venus and Mercury are most easily seen with the unaided eye as "stars" in the west after sunset or in the east before sunrise because they are orbiting close to the Sun. In the middle of the night,

we are looking away from the Sun and therefore away from Venus and Mercury. See the diagram below. [Figure 8.27] [CC 1: Patterns]

Figure 8.27. Orbits of Earth and Venus

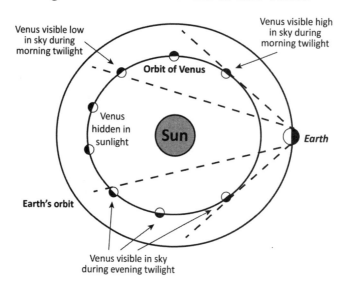

Application Question

On June 30, 2007, Trent took a long-exposure photo looking west out of his bedroom window at 9:30 p.m. On it he marked the positions of Saturn, Venus, and some nearby constellations (Leo and Cancer) (see Figure 8.28). If he took another photo looking out the *same* window at the *same* time on June 30, 2008 (one year later), which of these four objects would Trent expect to see, and why?

Figure 8.28. Constellations

> *Model response:* Trent would see the two constellations Leo and Cancer, because the Sun and Earth would be in the same positions relative to the stars. However, the two planets in the photo, Saturn and Venus, each orbit the Sun at different speeds. It takes them different amounts of time to orbit the Sun, so after one year they

would not be in the same position relative to the Sun and Earth. [SEP 6: Constructing Explanations and Designing Solutions; CC 1: Patterns]

Common Beliefs

Indicate whether the statements are true (T) or false (F), and explain why you think so (*model responses shown in italics*).

1. In general, the Moon rises in the evening and sets in the morning. *(F) The Moon rises in the east and sets in the west, because Earth is rotating on its axis. However, when the Moon rises and sets depends on the relative positions of the Sun, the Moon, and Earth. A new Moon rises in the morning and sets in the evening.*

2. When the Moon is new, the Sun is not shining on it. *(F) The Sun is always shining on the Moon except in the rare cases when there is a lunar eclipse when Earth is directly between the Sun and the Moon. This happens when the Moon is full. When the Moon is new, it is between the Sun and Earth. As usual, half of the Moon is lit, but most of the lit side is turned away from Earth.*

3. The time it takes to go from full Moon to full Moon is the same time it takes the Moon to orbit Earth one time. *(T) The Moon takes slightly longer than 28 days to orbit Earth. If the Moon is positioned relative to Earth so as to appear full, 28 days later it will be in the same relative position and will again appear to be full.*

4. There is one lunar and one solar eclipse approximately every six months. *(T) Both a solar and a lunar eclipse occur approximately every six months, when the orbit of the Moon crosses Earth's orbit.*

5. A lunar eclipse occurs when Earth moves into the Moon's shadow. *(F) A lunar eclipse occurs when the full Moon moves into the shadow of Earth. A solar eclipse occurs when the new Moon moves in front of the Sun and blocks all or part of its light, casting a shadow on Earth.*

6. An eclipse can be "total" for only a few minutes. *(F) During a solar eclipse, totality lasts only a few minutes, because the Moon's shadow is small and covers only a small part of Earth. During a lunar eclipse, totality can last an hour or more, because the shadow of Earth is large compared to the Moon.*

7. If Venus takes 225 days to orbit the Sun, then we will see Venus in about the same place in the sky approximately every 225 days. *(F) In 225 days, Venus will be back in the same position relative to our original position, but we will have moved partway around our orbit, so our position relative to Venus will be different.*

8. If a planet is seen in a certain constellation tonight, it will most likely still be in that same constellation at this same time next year. *(F) In a year, we will be back in the same*

position looking at the same constellations in the night sky. However, the planet is orbiting the Sun at a different speed and will be in a different position relative to ours.

9. All planets are spherical; therefore, when viewed from Earth through a telescope, planets appear as round disks. *(F) Planets do not generate their own light; therefore, we see only the portion of them lit by the Sun. That might not be the entire side facing Earth, so the planet may appear as a partial disc.*

9

MODIFYING AND DESIGNING YOUR OWN PROBLEMS

Chapters 5–8 presented 13 problems directly addressing four different Earth and space science (ESS) content strands. These problems are a good way to introduce you and your students to problem-based learning (PBL), and we hope the lessons are useful in your classroom. But these problems are by no means enough to cover every concept in your curriculum. There are more science ideas than we can fit in this book that would make excellent PBL topics.

So you may want to consider developing your own PBL lessons by modifying the ones we have presented, revising a current lesson plan to fit the PBL structure, or maybe even creating your own PBL problem from scratch. In this chapter, we will share some strategies for writing PBL lessons and tips for creating problems that are rich, engaging, and ideal for addressing the standards you need to teach. We have created a series of steps to think about as you design your own PBL lessons. The tips we provide are based on our own experiences and the advice of classroom teachers who have successfully implemented their own PBL problems. The process we recommend also follows the principles of backward design described by Wiggins and McTighe (2005).

Selecting a Topic for a Story

One of the features of a PBL lesson that distinguishes this format from some other approaches is the importance of the context in which the problem is situated. The story, consisting of Page 1 and Page 2, is designed to engage students in the concepts of the lesson, help them recognize the practical relevance of the concept, and give them a reason to want to pursue the challenge and find a resolution to the problem. If you elect to develop your own PBL lessons, one of the issues you will need to address is how to write a story that will encompass these important features of a good PBL problem. So where do you begin?

For many teachers, especially those who plan to use the principles of backward design, the first step is usually to identify learning targets, align the lesson with standards, and write objectives. Although these are critical steps, there is an important place for inspiration as a starting point in the planning as well. As we discuss how to identify the story, we'll start with the inspiration or creative idea and then move toward identifying standards and learning goals as part of the process as well. In actual practice, the two ideas—the creative inspiration and the methodical planning for standards—are closely tied. We encourage you to seek a balance between the two.

CHAPTER 9

Finding Inspiration for an Authentic Context

For many of the problems included in this book, the idea for the story was inspired by a real event. Some were written to reflect real events or a local problem or phenomenon that could be used to spark a science lesson. In other cases, the authors built a fictional—but realistic—story based on their understanding of the science concept and its practical implications. In developing your own problems, you will probably find that the best ideas share this authentic character. A real or authentic problem needs to be believable or even based on a real problem!

That inspiration can come from several different sources. Sometimes a newspaper article, an item on a TV news broadcast, or a story you hear about on the radio may be the spark that starts a new PBL problem, such as a story of sand being replenished at a local beach after storm erosion or a story about the formation of a sinkhole that destroyed a street or house. Or maybe a real problem in your community, such as a controversy about new restrictions on the digging of drainage ditches that feed into a local river, might be the inspiration. If your class works with a scientist from a business or university in your area, discussions with the expert in the field may lead to questions that lend themselves to the PBL format.

One valuable source of inspiration can be the questions students ask or their interests in science. Your students, especially in younger grades, are curious about the world around them, and they may bring ideas to class that would be excellent PBL problems. When a student brings a fossil to class and asks what it is or asks about last night's story on the news about a lunar eclipse, you may be able to build your lesson from these questions. One advantage to this approach is that you can be certain that at least some students already want to know the answer before you start!

There are strategies you can use to help elicit students' interests and questions in the classroom. One is to ask students in the first days of your class to write down some "I wonder … " statements. These are short written comments about science questions they would like to explore. You can sort through these responses and find patterns in their interests or maybe find a few outstanding questions you could develop into a PBL lesson.

One kindergarten teacher in the PBL Project for Teachers even created a Science Questions bulletin board in the classroom. During her weekly science lessons, any questions students raised could be written on a note card and thumbtacked to the bulletin board. She would then use those questions to design investigations for later lessons.

In other situations, you may need to build the contextualized problem. If you've already identified a standard and connected the concept to a context, you might be able to write a fictional but realistic scenario that presents a problem you can relate to your PBL plan. The astronomy problems about solar and lunar eclipses (see Chapter 8) are examples of this process.

You may also find it necessary to combine some of these sources. One group of teachers in the PBL Project for Teachers designed a problem about the increasing growth of poison ivy in the Midwest for seventh- and eighth-grade science classes (Boersma, Ballor, and Graeber 2008). The original idea came from an online article that mentioned that poison ivy is growing faster and larger because of increasing levels of carbon dioxide (CO_2) in the atmosphere. The story addressed science concepts that fit perfectly in the eighth-grade standards about human impacts on the carbon cycle and the impact of humans on climate.

To teach this lesson about poison ivy, Kylie (the same teacher described in Chapter 4, pp. 52–59) modified the problem for the students in her small rural community where deer hunting is very popular. She decided to make the story about a student and her father going hunting and noticing that the poison ivy plants were larger than usual and found in more places.

Once she had identified her learning goals and an authentic problem, the rest of the process was easy. In the next sections, we will refer to Kylie's PBL lesson again to see how she created the elements she needed for her lesson.

Addressing Standards

Even though you may begin your PBL planning with a story or context that inspires your idea, at some point you need to make sure that the lessons you develop address the standards that teachers are expected to cover. We suggest you follow the inspiration with some thought about your content standards.

Because you must account for students' learning of specific concepts, the content standards may be a good place to start planning. For some science concepts, you may have units in which you have strong inquiry-based activities and a plan that works very well for your students. A good way to start your new PBL plan might be to identify a unit that you feel needs to be revitalized. Or you may wish to choose a standard that you already teach well but that you want to update or revise.

When you think you have a context or story that would fit in your curriculum, the next question should be whether the story actually addresses appropriate learning goals. For instance, if the story you select is based on a newspaper article about a street destroyed by a sinkhole, you can search for standards that focus on erosion, landforms, and the potential impact of human activity on the environment. If there are standards that would include this concept, you have a story that fits your needs, and you can continue your lesson development. This topic also has the advantage of including other examples from different contexts, such as a landslide or siltification of a river (*siltification* is the rapid buildup of silt, usually behind a dam). For students to fully engage in the PBL problem, the story needs to be something they can relate to on a personal level, but teaching appropriate content is your primary responsibility.

Writing Assessments

In the backward design model (Wiggins and McTighe 2005), the next step after identifying a learning goal or objectives is to consider how to assess student learning. We suggest thinking about this before developing a story for the PBL problem you are working on, because knowing what student learning should look like will help you focus on the necessary questions and information for guiding students toward the goal.

Format of Questions and Assessments

Just as in any science unit, there are many ways to assess student learning. A test or a project is an effective way to assess certain types of learning, and these may be appropriate instruments for your PBL lesson. But for now, we will focus our attention on some of the types of questions and assessments described earlier in this book. We have found that certain types of assessment questions seem to be most effective in revealing students' thinking about the problems teachers present in a PBL lesson. Let's take a quick look at the types of assessments you might choose to write. For more detailed examples, please refer to Chapter 4.

GENERAL QUESTIONS

A general question is an open-ended response item that simply asks for a student's overall understanding of a concept. This type of question is very well suited for a pre/post assessment. Let's look at an example that might work for Kylie's poison ivy problem: "How would changes in CO_2 levels affect the growth of plants?"

Although this type of question is good at identifying big gaps or fundamental misconceptions, no one question can tell a teacher everything about students' thinking. The general questions featured in Chapters 5–8 are helpful in assessing students' overall understanding and their ability to state connections between ideas. But most students' responses to these questions may cover several ideas at a very shallow level, especially if students perceive a time limit in answering the questions. They often try to touch on as many related ideas as possible, but they do not give in-depth explanations. The teacher should remember that general questions like this example are not designed to probe or tease out the specific details that might be a barrier to understanding science concepts.

APPLICATION QUESTIONS

Although the general questions give you a glimpse into students' thinking, teachers still need a way to assess the depth of their understanding. In the PBL Project for Teachers, we explored some possible assessment strategies and eventually developed a structure for questions that ask learners to apply their ideas to authentic problems. The application question lets you ask an open-response question that elicits a slightly deeper understanding of a specific concept. The format for the question that gives the most information describes a phenomenon, asks for an explanation of how or why the event happens, and usually includes some key terms or ideas that should be included in the response. Here is an example for the poison ivy problem:

> NASA has investigated the use of plants as a source of food and oxygen in long-term space flight. They've found that plants grow faster in enclosed spaces if animals or bacteria are also growing in the same space. Explain why these plants grow faster when other kinds of organisms are present. Your response should include the materials plants need in order to carry out photosynthesis.

You can see that the question gives some hints about the type of information relating to the scenario when it states what the response "should include." This may seem like it gives away the answer, but if students do not have a deep understanding of the process of photosynthesis, their explanations will reveal which parts of the concept are unclear or confusing to them and which parts they simply have wrong.

Like the general questions, application questions can also be used in a pre/post assessment model. You may also want to include application questions if your goal is to have students understand the details of a concept, explain a concept in more than just general terms, or focus on a specific part of a more complex process. When used together, the general and application questions provide a coherent picture of what learners know.

TRANSFER TASKS

Transfer tasks are questions that present the learner with the same concept as the PBL problem, but in a different context. The goal of this type of assessment is to identify the learner's ability to apply the new concept to other situations. This transfer of understanding is an important indicator of deep understanding (Bransford and Schwartz 1999).

The transfer tasks developed in the PBL Project for Teachers were worded as open-response questions that gave enough detail to help the reader understand the new scenario. An example can be seen in Chapter 7 in the Water So Old problem:

> Write a story following a water molecule from its time in the primordial oceans before there was life on land to Joe's glass of water. Your story should include at least four of the places that Roberta and Joe talked about. You should also note when the water molecule was part of a process that transferred energy.

This example connects the question to the original concept of the water cycle presented in the problem. The question is flexible enough to work with multiple grade levels. As with any assessment, you should base the level of your question on the learning goals and level of complexity you expect in the responses.

PROBLEM SOLUTION SUMMARIES

One way to determine if each group or individual student has learned the target concept is to ask for a written summary of the solution to the problem. The summaries elicit a coherent explanation of what solution the student feels will resolve the problem or challenge and why that explanation makes sense. Like the previous examples, this is an open-response question. See Chapter 4 (pp. 59–60) for an example of a solution summary.

Writing Model Responses

Once you've written assessment questions for your PBL problem, an important strategy is to write model responses. Think of this as the "answer key" to the assessment or as an ideal answer. The model response represents the type of complete and accurate answer you hope your students will be able to provide at the end of the lesson.

Writing this response early in the development process will help you focus your planning on tasks that help learners move toward successful achievement of the learning goals you have identified in the first part of your lesson plan. Some teachers will highlight key elements of the model response or mark those elements with a boldface font to help organize their evaluation of student responses. If you are using the model response to assign points for a grade, this can also help you identify the points you award for an answer.

Chapters 5–8 included many examples of model responses written for each of the problems. Use these as examples as you write your own assessments.

Writing the Story
Writing Page 1

With a story context in mind, the learning goals identified, and a plan for assessing learning, it is time to write the story. In the model we adopted from partners in the field of medical education, the story consists of Page 1 and Page 2. Page 1 is the initial introduction, and Page 2 is presented to learners after analyzing Page 1 and generating lists of "known" information, questions, and hypotheses based on the story.

When you read examples of the stories from Chapters 5–8, they look very simple to write. They are short, and they include limited information. But in practice, the process of writing the story can prove more challenging than it looks!

One of the challenges, especially with Page 1, is determining how much information to leave out and how much extraneous information to include. Remember that the PBL story needs to be "ill-defined" in that information is missing. Some information is provided that will not help in finding a solution. Most teachers instinctively want to provide all the needed information. It is easy to give more than learners truly need. But if you leave out too much information, students may not connect the context to the learning goals of the lesson. Finding a balance takes some careful thought and editing.

For Page 1, try to give just enough information to grab the attention of learners. Page 1 of the story needs to describe the context and present some kind of question or challenge that needs to be answered. Let's take a look at an example from Chapter 8, the Page 1 story from the Copper Moon problem. In this problem, students explore what causes a lunar eclipse, and why two people located far apart will see the same event at the same time.

Page 1: The Story

Copper Moon

Lakesha had moved from Florida to Oregon the previous summer but still missed her friends in Florida. One night her phone rang while she was sleeping. She grabbed it, wondering who would be calling at 3:30 a.m. It turned out to be Miranda, one of her friends from Florida.

"You're never going to believe this!" Miranda screamed from the other end of the line. "Look outside your window!"

"Miranda," Lakesha cried, "it must be 6:30 a.m. for you. What's going on?"

"Everything's OK. I've been up all night working on my robot for class. But I know you are so interested in eclipses, so I thought you'd like to see another one! I've been watching this one for an hour. I'm looking at the Moon right now. It's totally eclipsed. It's gorgeous! Look out the window!"

Lakesha muffled a long yawn. "What am I going to do with you, girl?" Then she laughed. "OK. Let me take a look." She padded barefoot across her room to her window. And there, high in the southwest, she saw a copper Moon in a velvet sky. They watched together for a few minutes, speaking softly, until Miranda had to go to sleep. But Lakesha leisurely enjoyed the rest of the show.

Your Challenge: *Use a model to explain why it was possible that Lakesha and Miranda could leisurely watch the same lunar eclipse together, but not a solar eclipse.*

In this story, the boxed text gives a context for the story. The shaded text gives some information that may be relevant to the challenge. The underlined text states the challenge presented to students. The rest of the story gives details that are not essential to solving the problem.

Teachers often naturally have the urge to write more details into the story, but keep in mind that Page 1 is intended to *start* the conversation, not answer all of the questions. One of the things teachers are tempted to do is to define terms in the story. In this case, students may not wonder why the times are different for the two girls, but the structure of the PBL analysis framework gives learners a chance to list "Where are Lakesha and Miranda?" as a "need to know" issue. Even if terms are introduced that you suspect students won't know, avoid adding a definition. In Page 1, you only need to spark some curiosity, and students will let you know when they need a definition or explanation.

Writing Page 2

Page 2 of the story provides a bit more information, as shown in the Copper Moon example.

Page 2: More Information

Copper Moon

On September 6, 1979, there was a total lunar eclipse. Observers would have been able to see the lunar eclipse without special equipment, as long as the sky was clear in their location.

It was *visible to most observers in the Western Hemisphere*. The *eclipse began at 5:18 a.m. Eastern Daylight Time (EDT)* when the *Moon moved into Earth's shadow*. **Many observers report that the Moon looks bigger or "more three-dimensional" once an eclipse begins.** On September 6, 1979, the Moon was *totally eclipsed when it moved through the darkest part of Earth's shadow, which occurred between 6:32 and 7:18 a.m. EDT*. The eclipse was completely **over when the Moon exited Earth's shadow** at 8:30 a.m. EDT. The later part of the eclipse was not visible to Miranda in Florida.

This new page in the story adds some information to help the learners move closer to a solution for the problem, but it still leaves out some essential ideas. The italic text directly answers questions that are likely to arise during the Page 1 discussion, including "Can an eclipse be seen by everyone on Earth, or only by people in one area?" and "How long does a lunar eclipse last?" But there will likely be other "need to know" items that Page 2 does not answer. It is important to avoid giving all of the information needed to solve the problem, because the research phase helps students develop important skills, too.

The bold sentences are new information that may offer information to help students visualize an eclipse. The other text (roman type) includes interesting bits of information that may help students understand lunar eclipses but might not be needed to help students reach the ultimate learning goal. Having both relevant and non-essential new information on Page 2 is important. Remember that identifying useful and extraneous information is a part of the process of solving real-world problems. Your students need to practice this skill.

If the two pages of the story are written well, your students will have enough hints to keep their focus on the relevant science concepts, but they will still need to think about new questions that will require further research. As you gain more experience facilitating PBL lessons, your sense of how much information to include or leave out will become keener, and the initial phases of writing a problem will become easier.

Getting Feedback

When you have a draft of a problem that seems appropriate, we suggest asking someone else to read it. If possible, have a nonscientist or someone who is not an expert in that particular science concept read the draft. This type of review lets you see from a new perspective whether the story is engaging and whether the information included is too much or too little. Ask your reviewer to tell you what science concept he or she would use to solve the problem. The answer may suggest that the story needs a bit more information to start your learners down the path you intended. If the reviewer can tell you the entire solution just from the Page 1 story, you may have provided too much information, assuming your reviewer is not already familiar with that particular story or context.

This review is an extremely valuable step! Once you have more experience writing problems, you might elect to skip this step, but the information you get from this review is so important that we suggest you always have another person read your story as you develop it.

Tips for Writing the Story

As you write, try to avoid writing the problem like a textbook question. The language you use should be comfortable and accessible to students and should sound more like a newspaper article or a conversational story. One strategy is to make the story a dialogue between characters. You may also be able to adapt "cases" from other sources (Rose, Schomaker, and Marsteller 2015) to fit within the PBL framework. You can use the problem stories in Chapters 5–8 as examples, and you will find several "styles" represented in those examples. And it is fine to have some fun with the story!

But the writer also needs to work to make the story believable. The ideal story is a true story. The story in the Lassen's Lessons problem in Chapter 6 is taken from the travels of one of the content experts. In most cases, these are extremely easy stories to write. Others have been fictionalized, but they are real problems. An example is "Obsidian Sun" in Chapter 8: The characters in the story are fictional, but the questions are real, and the story could very well happen in the real world, including data from a historically accurate eclipse. Several other problems in this book are also fictionalized. The key is to not make

Box 9.1. Tips for Writing a PBL Story

- Use language that is accessible to students.

- Make the story as realistic as possible, even if it is fictionalized.

- Page 2 should answer some but not all of the "need to know" issues from Page 1.

- Have a nonscientist review the story. Ask if he or she can tell you what problem the learners need to solve and a possible solution.

  o If the person can solve the problem without other sources, you've given too much information.

  o If the person cannot identify the problem or connect to some broad science topic, you have not provided enough information.

- Add some color to the story with information that is not directly connected with the story, but not too much.

- Keep each page of the story short.

up a story that is beyond the realm of possibility. The more realistic, the better! (Box 9.1 gives more tips for writing a PBL story.)

Integrating Investigations

Sometimes students will identify "need to know" issues that could be addressed with a hands-on, inquiry-oriented investigation, such as a question about what happens to the barometric pressure as air is heated by the Sun. As you plan your own PBL problems, you will likely face the question of whether to plan to include one of your existing lab activities as a part of the lesson plan. Investigations are certainly an essential part of the science curriculum, and they have a place in the PBL framework. However, there may be a need to modify how the investigation is presented to students. In this section, we will suggest considerations for fitting a lab activity into your PBL problem.

The first consideration is the fit of the activity with the learning objectives of the PBL lesson. Early in your planning, you identified the desired outcomes. Does the lab you'd like to include in the plan directly address those outcomes? If not, a better place for the investigation might be before or after the PBL problem. If the lab does contribute to students' learning of the target concepts, then it might be an appropriate strategy.

For instance, you might have a very good lab activity in which students test rock samples using hydrochloric acid (HCl) to see if they are made of carbonates like limestone. Would it fit during the geology problems about plate tectonics in Chapter 6? Probably not! The central concepts of those labs are about the rock cycle, not qualitative tests for minerals. The HCl test is valuable, but would be more appropriate in another part of the ESS curriculum.

If the activity you are considering does fit the goals of your lesson, you should also think about how the lab is presented to students. In many lab activities, the preliminary materials and the handouts with the lab will give much of the information students might need to complete the investigation. Many of these labs are described as "confirmation" labs (Bell, Smetana, and Binns 2005). To fit within the framework of a PBL problem, the lab activity should provide some important piece of information that will fit within the entire collection of evidence students gather. A lab activity will not replace the research phase but will complement it.

So examine the lab handout. Is there a specific question presented as the focus of the investigation? Does it include follow-up information that gives away too much about your problem too soon? If so, revise the handout. You can remove some of the background information or replace it with your own version that fits within the context of the story you have written. You can also write a testable question that narrows the students' focus to the evidence you need them to gather for the PBL problem. Any follow-up questions should be centered on having students relate the data and evidence to the problem.

Taking time to ask questions about the fit of the lab to your PBL problem is very important. If the investigation is only tangentially related to the learning goals, the lab will be a distraction, and students may lose momentum in their drive to solve the problem. If the lab handouts overtly present the answer to your problem, the investigation negates the need to do further research, or it may remove the opportunity for students to construct their own solutions. All of these are critical elements in learning science and should not be skipped.

The following tips summarize things to think about when integrating an investigation or hands-on lab as part of a PBL lesson plan:

- Check to ensure the lab addresses your learning goals.

- Remove excess background and follow-up information.

- Write a testable question that identifies the purpose of the lab.

- Let the investigation focus on *one part* of the evidence needed to solve the problem.

Identifying Potential Resources for Students

When the story is written, one of the next steps is to look for sources of information that your students will need to find in the research phase of the PBL process. In this stage, students will search the internet, textbooks, library materials, and any sources you provide for information that will answer the "need to know" items they feel are most relevant to the problem.

You have different options for providing those resources or access to the resources. Many teachers/facilitators let learners search the internet in class on tablets or laptops or in a computer lab. There are many search engines that you can use to find sources of information for most authentic problems.

However, students may find that there are so many resources that they are overwhelmed by the sheer volume of content they need to sort through. If your students are not ready or you don't have time to discuss strategies for narrowing their search terms or evaluating the strengths and weaknesses of different sources, you may find it helpful to locate several sources and provide a set of links for students to use (see Box 9.2). If your school has a course management system or a course website that students can access, you can post links for students to choose. Experience tells us that you can provide more links than learners need, and your students can begin to practice strategies for selecting sources from a list.

In some classrooms, or with some students (especially younger learners), teachers may not be comfortable sending the class to the internet to do a search. In these cases, you can make copies of some sources from websites, newspapers, magazines, books, and other texts. Make sure you comply with copyright and fair use practices with this option. With a set of resources in hand, you can produce packets of related information.

If you provide the research materials, there are a couple of useful strategies for sharing the information with student groups. One is to provide a packet with all the materials for each group and allow them to select the files that are more important for their research. This is an especially effective approach if you have asked each group to develop their own solution and share their final answers with the class. Each group will receive all the text materials, so they will have the necessary information to build a solution.

For a lesson in which each group selects different "need to know" issues to research, the teacher can sort files by topic and give each group a different packet. The advantage in this strategy is that the groups will need to share their findings with each other to enable the entire class to construct a solution. This process models the work of real scientists and helps build a sense of teamwork in the classroom.

Whichever strategy you decide to use for the research phase of your lesson, you need to make sure there are relevant sources available to students. Take the time to do searches using terms you think students are likely to use, and check the sources to make sure they are reliable and scientifically accurate. One way to evaluate the trustworthiness of the source is to check the author's credentials. If he or she is a qualified authority, the source is more likely to be accurate. But remember that just about anyone can post a website as a "source," so not all the content you find will list an "expert" as the author.

You may also want to check the URL. Web addresses that end in *.edu* and *.gov* are usually more reliable, although many *.com, .org,* and *.net* sources also contain excellent information. Sites posted by nonprofit organizations like nature centers, conservation groups, and citizen science projects will have a *.org* suffix, but they can be among the best sources.

When you do this search, you should use a computer in your classroom or school. If you find a source at home, you may find that students cannot access the site if your school has a proxy server or a filter that blocks some websites. By locating sources in the same location where your students will conduct their searches, you can make sure the resources will work for your class when they need them. Identifying relevant sources in advance will also let you submit a request to your network administrators to give permission in the filter for the sources you need for a given lesson.

Printing copies of the sources you find may also be a good idea even if you plan to let students do their own internet searches. Technology can have a glitch at any moment, and if the server hosting the site you want students to visit is not working, or if the organization moves the site or redesigns the page, your students might run into broken links or search lists that lead to a different page. Be prepared for the possibility of a technology glitch!

Box 9.2. Tips for Locating Sources

- Use multiple search terms that students may select.
- Check sources to see if they are reliable and accurate.
- Print backup copies in case links are not working.
- Search from the classroom to see what your school's filter will allow.
- Submit in advance a filter exception for any sites students need to use.
- Keep a list of links available on a course website in case you need it.

Writing the Solution

Just as when you write your assessment questions, one important step is writing a model response for the solution to the problem. It can be easy to overlook this step as you plan your PBL lesson. Having a solution written before presenting the problem to students will help you plan to evaluate the final solutions your students produce.

But your solution needs to be flexible, or you may even need to include more than one possible answer. Some PBL problems may have more than one acceptable solution. In most of these cases, students may have different sets of values they use when selecting a solution. For instance, any problem relating to a plan to use natural resources may lead to different choices. Students may choose to preserve a habitat or propose a responsible strategy for using a resource such as lumber or minerals; the choice depends on the degree to which they value conservation of the natural site versus balancing human and economic resources. Be prepared for the possibility of multiple solutions.

Writing a model solution can also help you ensure that the wording of the story, the available resources, and the challenge presented to students are designed to permit learners to reach an appropriate solution. You may find that you need to revise the story or locate additional materials to foster your students' successful completion of the lesson.

Piloting a Problem

Another strategy to try as you plan a PBL lesson is to pilot the problem with a small group of volunteers. Try to create a group that is not familiar with content and ask them to read the story. By letting them apply the PBL analytical framework (What do we know? What do we need to know? and Hypotheses?), you can find potential sources of problems your students would encounter. As with having a reviewer read the story, a pilot group can be a great source of feedback to catch issues you might overlook.

Some possible groups from which you can build a pilot group include a science club or other student group, family members, or fellow teachers in other subjects. If your school has created professional learning communities (PLCs) for school improvement, your PLC group may be an ideal place to share your lesson ideas, including a request to pilot the problem.

The pilot group will be better able to provide useful feedback if you give them questions to consider as they review the problem. The PBL Lesson Talk-Through at the end of this chapter (also available on the book's online Extras page at *www.nsta.org/pbl-earth-space*) can serve as a feedback guide for the pilot group participants.

Modifying Existing Lessons

If you are like most teachers, you probably have a very full curriculum plan, and adding new lessons to the schedule will only work if you drop another lesson. You probably also have lessons or activities in your files that you really like but could be updated or modified. PBL may be the format that will let you revise existing lessons. Modifying what you have is a great way to keep the ideas that work while still implementing new strategies.

So think about your current set of lesson plans. Do you have a lab activity, assessment question, or discussion starter that you feel could be more engaging? Does it relate to a real-world problem that your students can relate to? If so, these are great opportunities to modify what you have! Let's look at an example of a way to adapt your current activities.

In many ESS classes, teachers include an activity in which students build models to show the movement of Earth's crust. This is a process that is hard to see in real time, so models allow students to simulate plate tectonics in a shorter time frame. The models might include online simulations and tutorials, but many teachers choose to make physical models with crackers, candies, frosting, plastic wrap, or paper plates. You can see some examples of the use of physical models in the Chapter 6 problems. The models are a good way to show students how the process of plate tectonics leads to certain types of formations depending on the type of plate boundary. But unless students have a reason to use the examples they have built, the concepts may not be robust or lasting. The three problems in Chapter 6 each provide a context that makes understanding the models more relevant.

For the Chapter 6 problems, the process of modifying the existing lesson began with brainstorming some potential contexts that can demonstrate different types of plate boundaries. Although there are several possible examples, the authors of these problems chose the Upper Peninsula of Michigan and two locations in California because their students could relate to the examples in those areas.

The authors then drafted stories that set the scene that students will think about as they explore the geologic history of the three locations. They shared the stories with colleagues to get feedback, made some revisions, and wrote assessment questions. The lessons still leave an opportunity for an investigation with plate tectonics models using the existing lab activities, but the topic is now presented as a PBL problem.

Your own curriculum plan is a potential source for PBL ideas. Take the time to check your current set of lessons and see if you can find a place to be creative in writing a new PBL problem!

While we are thinking about modifying lessons, please remember that problems are often specific to a particular place and time. Chapters 5–8 include lessons that were written for learners in Michigan. Some of the local issues that led to story ideas may not be as engaging for your classroom. So feel free to change the stories if you find it helpful! Modify what we have presented, just as you may choose to modify your own lessons.

Resources for Writing PBL Problems

At the end of this chapter are two aids that you can use as you plan your own lessons. The first is a PBL Lesson Template with prompts to help you think about the parts of the PBL lesson structure we have presented. The second is a PBL Lesson Talk-Through, which includes a list of questions you should think about and discuss as you plan and revise your lesson. Both of these documents are also available on the book's online Extras page at *www. nsta.org/pbl-earth-space*.

References

Bell, R. L., L. Smetana, and I. Binns. 2005. Simplifying inquiry instruction. *The Science Teacher* 72 (7): 30–33.

Boersma, K., L. Ballor, and J. Graeber. 2008. Poison ivy: Teachers designing a problem-based science unit. *MSTA Journal* 53 (1): 28–31.

Bransford, J. D., and D. L. Schwartz. 1999. Rethinking transfer: A simple proposal with multiple implications. *Review of Research in Education* 24: 61–100.

Rose, J., B. Schomaker, and P. Marsteller. 2015. CASES Online. Emory University. *www.cse.emory.edu/cases*.

Wiggins, G. P., and J. McTighe. 2005. *Understanding by design*. Alexandria, VA: Association for Supervision and Curriculum Development.

PBL Lesson Template

Authors:

Science Content Topic:

Grade Level:

Expected Timeline:

"Big Idea":

Standards:

Story—Page 1: *(narrative giving context, just enough to identify the challenge)*

Story—Page 2: *(additional necessary information, more details on the story)*

Potential Sources:

Assessment

Pre/Post Assessments:

Transfer Task:

Solution/Presentations:

Informal Assessments:

PBL Lesson Talk-Through

Name of Lesson:

Talk-Through From Student Perspective

Pretend to be one of your students. Without thinking about the answer or solution, answer the following questions:

- How would you interpret the task/challenge? What is the problem asking you to do?

- What questions come to mind that might help you solve the problem?

- How would you go about solving this problem?

- What would you as a student know that would be useful in solving the problem?

- What similar tasks might you have experienced before?

- What is confusing about the problem?

Talk-Through From Teacher Perspective

Answer the following questions as the teacher/facilitator for the lesson:

- What would you expect your best students to do with this problem?

- What would you as a teacher expect your less able students to do with this problem?

- What materials will be required? What resources?

- Which resources might be difficult for your students to find?

- How long do you estimate this lesson will take?

IMAGE CREDITS

Chapter 1

Figure 1.1: NSTA Press

Chapter 3

Figure 3.1: Janet Eberhardt

Chapter 4

Figure 4.1: Tom J. McConnell

Figure 4.2: Tom J. McConnell

Chapter 5

Figure 5.1: Saffron Blaze, CC BY-SA 3.0. *www.flickr.com/photos/saffron_blaze.*

Figure 5.2: Reid Priedhorsky, Public Domain. *http://reidster.net/trips/maps/usgs-full-600.png.*

Figure 5.3: Joyce Parker

Figure 5.4: Nicolas Raymond, CC BY 3.0. *http://freestock.ca/wishing_well_%C3%83%C2%A2%C3%8B% C2%9C%C3%A2%C2%80%C2%A6_g103-sideling_hill_highway_%C3%83%C2%A2%C3%8B% C2%9C%C3%A2%C2%80%C2%A6_p5101.html.*

Figure 5.5: Josh Hallet, CC BY-SA 2.0. *www.flickr.com/photos/hyku/526266144.*

Figure 5.6: NOAA and National Science Foundation, Public Domain. *www.noaanews.noaa.gov/ stories2009/images/fire_smoke.jpg.*

Figure 5.7: Hermanturnip, CC BY 2.0. *www.flickr.com/photos/hermanturnip/3540834299.*

Figure 5.8: Brewbooks, CC BY-SA 2.0. *www.flickr.com/photos/brewbooks/281936925.*

Figure 5.9: Soil Science @ NCState, CC BY 2.0. *www.flickr.com/photos/soilscience/5140039955.*

Figure 5.10: Kiran Foster, CC BY 2.0. *www.flickr.com/photos/rueful/5914598717.*

Figure 5.11: Johndal, CC BY 2.0. *www.flickr.com/photos/johndal/2835287487.* (flood); NASA, Public Domain. *http://earth.jsc.nasa.gov/SearchPhotos/photo.pl?mission=ISS030&roll=E&frame=111318* (river delta); Thomas Quine, CC BY 2.0. *www.flickr.com/photos/quinet/9690645555* (gravestone).

Figure 5.12: StooMathiesen, CC BY 2.0. *www.flickr.com/photos/stoo57/7118936423.*

Chapter 6

Figure 6.1: Joe Ross, CC BY-SA 2.0. *www.flickr.com/photos/joeross/6127862804/in/photolist-akuU7Y-54Yt4x-fMJk69-77WycF-781tZs-781tbq-77WzbT-5Di1LC-781sZs-781ssL-781skA-3hJvbC-77Wznn-781ujJ-2ABBNq-5xD7DU-akeBFG-fMJKFS-aktUiR-fMsbkK-NUpe4-77WzpF-54YwdE-8R4PET-fMrWfv-aktUDa-akvHUU-aksUTr-aks6Wx-akvJq7-agKWn1-3hBgnX-pWzfcy-781swh-77dLVV-*

IMAGE CREDITS

2AxdbM-fMJLKJ-3hFtq9-akcQXK-fMrZsg-akfD8G-akcR2K-fMJLrW-akbPSe-agH3Ua-5rWuA7-5baEzi-5xyvED-fMJx47-5xyGfP.

Figure 6.2: Joyce Parker (left); James St. John, CC BY-SA 2.0. *https://en.wikipedia.org/wiki/Copper_mining_in_Michigan#/media/File:Cupriferous_amygdaloidal_basalt_(Mesoproterozoic,_1.05-1.06_Ga;_Wolverine_Mine,_Kearsarge,_Upper_Peninsula_of_Michigan,_USA)_(17323753605).jpg* (right)

Figure 6.3: James St. John, CC BY-SA 2.0. *www.flickr.com/photos/jsjgeology/15713782910* (top); Joyce Parker (bottom)

Figure 6.4: Joyce Parker (top); jsj1771, CC BY 2.5. *www.everystockphoto.com/photo.php?imageId=21478300&searchId=ddcbbafd9c836941f6a8caab09524a2e&npos=15* (bottom).

Figure 6.5: Jim Gardner, CC BY 2.0. *www.flickr.com/photos/jamesthephotographer/745888834* (left); Lassen NPS, CC BY 2.0. *www.flickr.com/photos/lassennps/9688693194* (right).

Figure 6.6: U.S. Geological Survey, Public Domain. *https://commons.wikimedia.org/wiki/File:Lassen_Peak_Before_1914.jpg*.

Figure 6.7: Hampton (LassenNPS), CC BY 2.0. *www.flickr.com/photos/lassennps/11293045813*.

Figure 6.8: B.F. Loomis (LassenNPS), CC BY 2.0. *www.flickr.com/photos/lassennps/8435233161* (left); B.F. Loomis (Lassen NPS), CC BY 2.0. *https://en.wikipedia.org/wiki/Lassen_Peak#/media/File:Hot_Rock_and_Lassen_Peak_eruption_(8435233899).jpg* (right).

Figure 6.9: LassenNPS, CC BY 2.0. *www.flickr.com/photos/lassennps/8553662274/in/photolist-cZXACb-e2RKML-cxJwjj-cxJwQ1-cZXyFd-e2L5RR-bVp5LV-dffVT5-dffSEf-dffUU1-8xqMe-cDePAu-cDeLKJ-cLrMN9-cDeQGq-cZXzid-ouVdqj-dffUqd-oMo6yh-8rL6Tp-cDeKZY-oM8UzR-cMnipo-ouUVJz-oMq1Vi-cZXTmY-ouUXUr-futMP4-cZXzV7-cZXMdh-cZXGPN-5hUpL3-cZXUBU-cMnieN-6Zduav-cZXCGE-cZXFr5-6YnAT-6rN2QJ-5b7uMh-cDeLXU-43Cye-XsfeG-ccLmkq-54B7h-dffVuv-6Zdw1k-oKo4pd-o3RcWQ-6rHU4p*.

Figure 6.10: Glenn Scofield Williams, CC BY 2.0. *www.flickr.com/photos/glennwilliamspdx/3939615985/in/photolist-dRoMae-43Cye-icVEa2-icVDwD-718yzx-BuE83-4714r-x19Y2P* (left); LassenNPS, CC BY 2.0 *www.flickr.com/photos/lassennps/15302327265/in/photolist-pjdqLH-fuxG7t* (right).

Figure 6.11: James St. John, CC BY 2.0. *www.flickr.com/photos/jsjgeology/15590211992* (left); James St. John, CC BY 2.0 *https://commons.wikimedia.org/wiki/Category:Marin_County,_California#/media/File:Glaucophane-lawsonite_blueschist,_Marin_County_CA.jpg* (right).

Figure 6.12: Joyce Parker (left); Greg Willis, CC BY-SA 2.0. *www.flickr.com/photos/gregw66/5349510377* (right).

Figure 6.13: Zimbres, CC BY-SA 3.0. *https://en.wikipedia.org/wiki/Marin_Headlands#/media/File:Pillow_lava-Marin.JPG*.

Figure 6.14: James St. John, CC BY 2.0. *www.flickr.com/photos/jsjgeology/8513403566/in/photolist-dYiqh7-eCY3RJ-rsvjqz-dBM2dV-yDpxZ6-yTBAc1-ySvS8d-yDiYkJ-yUVx7s-pRzAJP-yWEnHK-xYT8c3-yVV6vp-xYQBQm-yDpZxk-xZ3vbp-yATZ3f-yDjyUd-yDpEMK-yDjeEb-yWHLHg-xWBBpD-yVW1cg-yTBRGj-yWHPwn-yDp6ND-yDieMY-yTvG7Z-yFLVSF-yDncT8-yTB2Vs-yAU9fU-ySvP8f-yWELBR-xYReHd-xYZ3TF-xWBEiV-yVYXJq-yDp38c-yTyKEL-yTvtx-R-yUT56h-yUSGPu-yZ6CY2-yTz14A-y2gtmL-yVT2C8-xYQLUo-y2qyrX-yAUAsw* (left); National Park Service, Public Domain. *www.nps.gov/goga/learn/education/graywacke-sandstone-faq.htm* (right).

Figure 6.15: James St. John, CC BY 2.0. *www.flickr.com/photos/jsjgeology/16638766949/in/photolist-68t3Jn-4riizn-Evw5QA-5Gj3EQ-rmj2Je-nRXBD5-qTJSAM-cJY7zS-34Ydj3-btBp4C-5KUfLs-btBods-d6HEVs-rCKShB-zteQW-rmre2G-dEGwhk-g9qVPv-i8923-p9cMJs-qFYxNp-5zqNq6-

8Xo8QF-zteR2-5FzPij-p9cMDh-9HHjn1-oWKg3z-jXyMt-p87uqu-g9pwhf-9HEtUB-uzosSG-jXz85-
9UDXb4-deHLxy-d6HEhL-jXyWU-7mgZn-egiqUb-oXbWuT-jXyBA-jXzhN-dfZoip-98dzdm-kqpovz-
jW5nN-GaWxs-jW5vG-kqppLa (left); Greenbelt Alliance, CC BY 2.0. *www.flickr.com/photos/
greenbeltalliance/775496776* (right).

Figure 6.16: Joyce Parker

Figure 6.17: Doc Searls, CC BY-SA 2.0. *www.flickr.com/photos/docsearls/15392616.*

Figure 6.18: David McSpadden, CC BY 2.0. *www.flickr.com/photos/familyclan/15583050608/in/
photolist-pK2d3m-oC3Nmb-7zHipp-6HwFfk-6JxSTt-6HbSFA-6HwGrr-6JqJ2E-6HwHEc-6HqqHw-
4NZasE-6JxP8g-ap2hjz-evhkvp-5xDGrd-7KGVPe-9s4Uc1-ap2hAK-8wVYzf-p2t2fC-dv3SQm-8gQJiL-
7aqzK7-dbHjd-dNHCpW-6JmzKz-6JBWKN-6Hqhkj-6Hqkks-6JxQmv-BWor8v-ecLHxT-bBkKhg-P8SF-
c5dkVA-9YieBV-kYjY1t-5jusqf-duXhht-dv3SV1-9Kk4zv-4gntt-anDFBq-dfXrai-7vmhAW-czYcBy-
MNxej-mL7oEg-56eon3-6HbUmL.*

Figure 6.19: Ben+Sam, CC BY-SA 2.0. *www.flickr.com/photos/wlscience/4834904545/in/photolist-xTXRfx-
8nigk9-9yVKwQ-8nfaGa-8nihxQ-8nf9Fc-8nij7d-8nijpG-8nfdnH-8nfbS2-8nfd7k-8nfcLp-yPRSZG.*

Figure 6.20: James St. John, CC BY 20. *www.flickr.com/photos/jsjgeology/16126500763/in/
photolist-qz3wQB-8C6P98-9eB2uR-3JKWuv-2phd3H-fbt1Xe-wCvEB-wuALD-8vdVAR-cEckyJ-
c1Bf4h-aF3su5-vJFiz-vJGxd-4rkyRC-7H7vcM-96pR17-ohYNRe-aNueCe-8LzwLA-7ArSgq-bN3tGc-
4H8SvE-8Svzbq-3oJYRn-aayE1L-9VSRFP-2EgKcJ-oR4c8v-2BK1cB-7jMa9w-6ivrCY-bGwb18-sbtAvr-
5HfS3J-FPL18t-8GMMGD-Wac6a-Wac4x-WfCFG-WfCHS-Waccp-WdWaQ-WfCRS-Wbpr8-WbEaZ-
7pSPis-Wac6R-WbEhV-aEi42G.*

Figure 6.21: kqedquest, CC BY-NC 2.0. *www.flickr.com/photos/kqedquest/1277970035* (top); Bagaball,
CC BY 2.0. *www.flickr.com/photos/bagaball/4262797265* (bottom).

Figure 6.22: Brewbrooks, CC BY-SA 2.0. *www.flickr.com/photos/brewbooks/397239263.*

Figure 6.23: Ekko, CC BY-SA 3.0. *https://upload.wikimedia.org/wikipedia/commons/e/ef/Limestone_on_
shale.jpg.*

Figure 6.24: Doug Becker, CC BY-SA 2.0. *https://commons.wikimedia.org/wiki/File:Sandstone_Boulder_
at_Maitland_bay,_Bouddi_National_Park_-_New_South_Wales,_Australia.jpg.*

Chapter 7

Figure 7.1: Greenland Travel, CC BY 2.0. *www.flickr.com/photos/greenlandtravel/15090503907.*

Figure 7.2: NASA/NOAA/GSFC/Suomi NPP/VIIRS/Norman Kurin, Public Domain. *www.nasa.
gov/centers/ames/news/2013/M13-39-nasa-west-summit.html#.VwkEmzYrKCR.*

Figure 7.3: Joyce Parker

Figures 7.4–7.5: Tom J. McConnell

Figure 7.6: Enid Martindale, CC BY 2.0. *www.flickr.com/photos/enidmartindale/7749342258/in/
photolist-csBEMG-wSmyu9-7C9xHC-8PPonW-4ZdZHg-cNMpth-pX8sut-813PQM-8F6XRY-
6zUCCB-qB1k34-fkUBdn-aG3Hkz-7R7aj6-bthbtd-7vvxM6-r7bG5-7L7ySx-pRTLqk-9wsDEK-
6P4GMW-8SrBeH-6NZy9F-oM8uua-9DR4aM-onXABa-bt2ZCJ-agoSu2-bv96VJ-pUhfVz-2HHf44-
aVaBw2-Lg54h-ooMgfm-viGaPj-7o9s8J-cmfgE-wLCqg-8XrErZ-9MG6Wy-qmXTff-5UHweP-bqqVaE-
4LargZ-jnYot9-7umnpS-7iYSW6-bNRZ8-agBodf-xJGNu.*

Figure 7.7: Ross Elliott, CC BY 2.0. *www.flickr.com/photos/ross_elliott/4779598557.*

Figure 7.8: Nikolay Dimitrov, CC BY 4.0. *www.e-cobo.com/stockphotos/fruits_vegetables/free_high_
resolution_fruits_vegetables_photo_155_5568_jpg.jpg.html.*

Figure 7.9: Dellex, CC BY-SA 4.0. *https://upload.wikimedia.org/wikipedia/commons/c/c9/ Pachycephalosaurus_Modell.JPG.*

Figure 7.10: Tom J. McConnell

Figure 7.11: Jérome Decq, CC BY 2.0. *www.flickr.com/photos/lesphotosdejerome/8405843005.*

Figure 7.12: Tom J. McConnell

Figure 7.13: Erica Annie, CC BY 2.0. *www.flickr.com/photos/fantasticalnature/18711562465.*

Figures 7.14–7.17: Tom J. McConnell

Figure 7.18: ravedave, CC BY-SA 3.0. *upload.wikimedia.org/wikipedia/commons/thumb/0/01/Example_ of_a_cold_front.svg/800px-Example_of_a_cold_front.svg.png* (left); ravedave, CC BY-SA 3.0 *https:// upload.wikimedia.org/wikipedia/commons/thumb/c/c4/Example_of_a_warm_front.svg/800px-Example_ of_a_warm_front.svg.png* (right).

Figures 7.19–7.20: Tom J. McConnell

Chapter 8

Figures 8.1–8.4: Tom J. McConnell

Figures 8.5–8.7: © NBC Universal Studios

Figures 8.8–8.9: Tom J. McConnell

Figure 8.10: Torbak Hopper, CC BY 2.0. *www.flickr.com/photos/gazeronly/7243724172.*

Figure 8.11: Ji Ruan, CC BY-NC-SA. *www.flickr.com/photos/jiruan/6486976185.*

Figures 8.12–8.19: Stellarium, Public Domain. *www.stellarium.org.*

Figure 8.20: Tom J. McConnell, adapted from *commons.wikimedia.org/wiki/File:Venus_Cycle.png.*

Figures 8.21–8.24: Tom J. McConnell

Figure 8.25: Zeynel Cebeci, CC BY-SA 4.0. *https://commons.wikimedia.org/w/index.php?curid=49689527.*

Figures 8.26–8.28: Tom J. McConnell

INDEX

Page numbers printed in **boldface** type indicate tables or figures.

INDEX

INDEX

INDEX

INDEX